041 889 21

Basic statistical quality control

Consulting Editor

C. T. Butler, B.Sc.(Eng.), C.Eng., M.I. Mech.E., M.I. Prod.E.

Head of the Department of Mechanical and Production Engineering
Nottingham Regional College of Technology

Also in the Series

Martin: Introduction to Feedback Systems

Basic statistical quality control

A. G. Hopper
Lecturer in Mathematics, Derby and District College of Technology

McGRAW-HILL LONDON
New York · Sydney · Toronto · Mexico · Johannesburg · Panama

Published by

McGRAW-HILL Publishing Company Limited
MAIDENHEAD · BERKSHIRE · ENGLAND

07 094109 2

PRINTED AND BOUND IN GREAT BRITAIN

Preface

This book has been written to meet the needs both of students requiring an introduction to statistics for engineering courses at technical colleges, and also of quality engineers in industry who require a basic knowledge of the fundamental techniques of statistical quality control.

No previous knowledge of the subject has been assumed and revision notes on pure mathematics have been included where appropriate. Mathematical argument has been reduced to a minimum and proofs etc., included as notes at the ends of chapters. Use of classical examples based on dice and playing cards has been avoided and the statistical theory has been related to practical engineering problems at the earliest possible opportunity in each section.

The material covered is sufficient to meet the requirements of the City and Guilds of London Institute syllabus in Statistical Quality Control, to Full Technological Certificate level, and to cover the statistics content of many other mathematics syllabuses for students in various branches of engineering. In addition, chapter 7 includes a description of Pre-Control Chart techniques, for the benefit of quality engineers in industry.

Numerous worked examples are included in the text and problems are set at the end of each chapter.

The author would like to thank his colleagues in the Mathematics Department at Derby and District College of Technology for their helpful advice and constructive criticism during the preparation of this work.

List of symbols

$\bar{x}$ Mean of sample variate.
μ Mean of population variate.
s Standard deviation of sample variate.
σ Standard deviation of population variate.
$\hat{s}$ Sample unbiased estimate of population standard deviation.
x_i The *i*th value of the variate x.
f_i The number of times (frequency) the *i*th value of x occurs in the sample (or finite population).
v Coefficient of variation.
n The number of *different* values of x in the sample (or finite population).
N The total number of members in the sample.
w Sample range.
p Simple element of probability (also used for 'proportion defective').
P Compound probability—a function of p and, perhaps, other variables.
q $1-p$.
Q $1-P$.
λ The average number of events per interval in a Poisson distribution.
z The standardized Normal variate, i.e., the variate z is Normally distributed about zero mean with unit variance.
r The suffix for a general term.

Contents

1. Probability

1.1 Introduction

Mention the word 'probability' and most people have a vague idea of what you are talking about. They think, perhaps, of horseracing and the 'odds' quoted by a bookmaker, based—at least in part—on the 'probability' that their particular choice of horse will win the race. Or again, they may think of gambling on the throw of dice, the 'odds' being based on the 'probability' of a 'winning' score coming up.

We all know that if an unbiased, six-sided die is rolled we are more likely to get an odd number than, say, a four, since there are three faces marked with odd numbers of spots but only one face with four spots. Thus an odd number has a greater 'probability' of appearing than does a four.

In order to be of any practical use 'probability' has to be defined in such a way that a numerical scale can be used and hence consistent measurements and comparisons of probabilities made. This will be developed in the course of this chapter.

Probability is of importance in engineering in that, among other things, it is the basis of statistical techniques such as Sampling Inspection and Control Charting. Some other applications are suggested in the problems set at the end of this chapter.

1.2 Empirical probability

Let us consider a box containing 1000 set-screw blanks (prior to thread rolling), of which 16 are defective in the following way:

	8	are defective in shank length only
	5	are defective in shank diameter only
	2	are defective in head thickness only
	1	is defective in shank diameter and length
Total	16	defective blanks

Supposing these blanks are well mixed, when one is taken out at random, it may be either defective or acceptable. If this one is put back and the box well shaken, then another one taken out, and so on, the whole cycle being repeated many times, a typical sequence of results might be as follows:

No. of trials (t)	No. of times a defective is taken out (d)	$\frac{d}{t}$
1	0	0
2	1	0·5
5	1	0·2
10	1	0·1
50	3	0·06
100	3	0·03
500	6	0·012
1000	15	0·015
2000	31	0·0155
		etc.

The fraction

$$\frac{\text{Number of defectives drawn}}{\text{Number of trials}}$$

is known as the empirical (or experimental) measure of the *probability* of drawing a defective blank from the box. If we consider the drawing of a defective blank to be a 'success', we arrive at the following definition.

Definition 1.

$$\text{Empirical Probability} = \frac{\text{Number of successes}}{\text{Total number of trials}}$$

1.3 Theoretical probability

Considering the same box of set-screw blanks, 16 of the total of 1000 being defective, we would say that there were 16 chances in 1000 of drawing a defective blank at a single trial. This fraction, 16/1000, is the theoretical probability of drawing a defective blank from the box

Definition 2. If a 'trial' can result in any one of n exhaustive, mutually exclusive, and equally likely events, and m of these events are regarded as 'successes', then the probability, denoted by p, that any one 'trial' will result in a 'success', is defined by

$$p = \frac{m}{n} \tag{1.1}$$

This definition contains three important phrases: 'exhaustive' means that there must be n and only n possible results of the trial; 'mutually exclusive' means that any event, once it has occurred, completes the trial and excludes all other possible events, for example, the drawing of a single blank which is defective in head thickness and the drawing of a single blank defective in length are mutually exclusive. However, the conditions 'defective in length' and 'defective in diameter' are not mutually exclusive since there is one piece which is defective in both of these dimensions. 'Equally likely' means just what it says: the events must be such that each has the same chance of happening as any other. For example, if two coins are spun there are four, *not three*, equally likely events possible (i.e., head–head, head–tail, tail–head, tail–tail), hence a head and a tail is twice as likely as two heads, or two tails.

Returning to our box of screw blanks this definition leads to the following probability figures:

Probability of drawing a blank defective in length only $= \frac{8}{1000} = 0{\cdot}008$
Probability of drawing a blank defective in diameter only $= \frac{5}{1000} = 0{\cdot}005$
Probability of drawing a blank defective in head thickness only $= \frac{2}{1000} = 0{\cdot}002$
Probability of drawing a blank defective in length and diameter $= \frac{1}{1000} = 0{\cdot}001$
Probability of drawing a defective blank $= \frac{16}{1000} = 0{\cdot}016$

When we say that the theoretical probability of drawing a defective blank is 0·016 we mean that in 1000 attempts we would *expect* to draw *about* 16 defectives; in 500 attempts *about* 8 defectives; in 10,000 attempts *about* 160 defectives and so on. It should be remembered that the emphasis is on the words 'expect' and 'about'. The larger the number of trials the closer the agreement we can hope for between theory and practice, i.e., as the number of trials increases the value of empirical probability becomes a more reliable estimate of the theoretical probability.

If we refer back to the table on page 2 we see that the value of empirical probability (d/t) came closer to the theoretical value of 0·016 (i.e., 16 defectives divided by 1000 blanks) as the number of trials (t) increased. It cannot be stressed too strongly that, although more trials *usually* means closer agreement with theory, this will not always happen. There is a theorem in statistics, known as Bernoulli's Theorem, which states that:

> As the number of trials increases, the likelihood that the difference between an empirical estimate of probability and its theoretical value will exceed any given value, decreases.

Since the number of 'successes' can vary only from zero to a maximum equal to the number of trials, it follows that 'probability' has a scale of values running from zero to unity. A probability of zero indicates that 'success' is impossible, while unity indicates that it is absolutely certain.

We will now consider two fundamental laws of probability. The proofs of these laws are set out at the end of the chapter.

1.4 The Addition Law of probability

Let us consider again our box of screw blanks. 8+5 (i.e., 13) blanks were defective in length or in diameter but not in both, hence, from the definition of probability:

Probability of drawing a blank which is defective on *either* length *or* diameter only—*not* both

$$= \frac{8+5}{1000}$$
$$= 0{\cdot}008+0{\cdot}005$$
$$= 0{\cdot}013$$

But the probability of drawing a blank which is defective on length only is 0·008 and on diameter only is 0·005, so the probability of drawing a blank which is defective in either length only or diameter only is the sum of the individual probabilities. This illustrates the following Addition Law.

Law 1—The Addition Law: If p_1 is the probability that a trial will produce event E_1 and p_2 is the probability that the trial will produce a mutually exclusive event E_2, then the probability p that the trial will result in event E_1 *or* event E_2 is given by

$$p = p_1 + p_2 \qquad (1.2)$$

If p is the probability that an event *will* happen, then it follows that the probability that it will not happen, denoted by q, is given by $p+q = 1$, since either the event does happen or it does not; there is no other alternative, so the sum of the probabilities must be unity (indicating certainty).

Hence
$$q = 1-p \qquad (1.3)$$

1.5 The Multiplication Law of probability

Now let us suppose that each of the 1000 blanks in the box is identified by a serial number. If a blank is drawn out at random (that is to say with no preference for one part of the box rather than another, so that every blank has the same chance of being picked) and its serial number noted, and it is then put back and the box shaken thoroughly before another is taken out for its serial number also to be noted, then there are 1000×1000 possible, and equally likely, sequences of serial numbers. Since 8 blanks are defective in length only and 5 in diameter only, it follows that 8×5 of the 1000×1000 sequences will correspond to a length-defective blank followed by a diameter-defective blank.

Hence the definition of probability gives:

The probability that two successive blanks, drawn from the box, will be one defective in length *followed by* one defective on diameter

$$= \frac{8 \times 5}{1000 \times 1000}$$
$$= 0{\cdot}008 \times 0{\cdot}005$$
$$= 0{\cdot}000\,040$$

i.e., the product of the separate probabilities. This illustrates the following Multiplication Law.

Law 2—*The Multiplication Law*: If p_1 is the probability that a trial will produce event E_1 and p_2 is the probability that a trial will produce an independent event E_2, then the probability p that two successive trials will produce event E_1 *followed by* event E_2 *in that order*, is given by

$$p = p_1 \times p_2 \tag{1.4}$$

Notice here that the emphasis is on *in that order*. If we require the probability p that two successive trials will produce between them, events E_1 and E_2, then we must use Laws 1 and 2 together, since we can achieve our objective by either E_1 followed by E_2 or vice versa.

Hence $\quad p = (p_1 \times p_2) + (p_2 \times p_1)$

i.e. $\quad p = 2p_1p_2 \qquad (1.5)$

Still thinking of our box of screw blanks, the probability that two successive drawings will produce one defective in diameter only and one defective in length only, in either sequence, is

$$2 \times 0{\cdot}000\,040 = 0{\cdot}000\,080$$

N.B. The algebraic proof of these laws is presented at the end of the chapter

1.6 Sampling—large batches

Up to now we have considered only single trials with replacement before the next trial. Next we must consider what happens when a small number of pieces is drawn at once from a large batch.

Suppose we have 10,000 blanks of which 160 are defective (the same proportion as before), and we draw a random sample of 5 blanks from the batch

Probability that the 1st blank is a defective $= \dfrac{160}{10000} = 0{\cdot}016$

Probability that the 2nd blank is a defective* $= \dfrac{160}{9999}$ or $\dfrac{159}{9999}$

$= 0{\cdot}016\,001\,6$ or $0{\cdot}015\,901\,5$

etc.

Probability that 5th blank is a defective* lies between 160/9996 and 156/9996, i.e., 0·016 006 4 and 0·015 606 2.

We see that this probability is very nearly constant over the sample, i.e., from first to last piece drawn.

When extracting small samples from large batches we can assume that the probability of extracting a defective remains constant for all pieces in the sample.

Our sample of 5 pieces may contain any number of defectives from 0 up to and including 5. We must now consider the probabilities of obtaining samples containing each of these numbers of defectives.

Let p = probability that any one piece is defective.

It follows from the definition that this is equal to the proportion of defectives in the batch.

The probability that any one piece drawn is *not* defective is q where $q = 1-p$ (see eq. 1.3).

In our case $p = 0{\cdot}016$ and $q = 0{\cdot}984$. Hence, by the multiplication law, the probability of 0 defectives, i.e., 5 good pieces $= q^5$

$= (0{\cdot}984)^5$

Similarly, the probability that the 1st is defective, and the remaining 4 good $= pq^4 = (0{\cdot}016)\times(0{\cdot}984)^4$.

The probability that the 2nd is defective and the remaining 4 good has the same value, as do all the remaining combinations of 1 defective and 4 good (i.e., ${}^5C_1$: see mathematical note at the end of the chapter). Hence, by the Addition Law, the probability of exactly 1 defective $= {}^5C_1 \times 0{\cdot}016 \times (0{\cdot}984)^4$. (The Addition Law is invoked since the 1 defective can be *either* the first *or* the second *or* the third, etc.) Similarly:

Probability of exactly 2 defectives $= {}^5C_2 \times (0{\cdot}016)^2 \times (0{\cdot}984)^3$
Probability of exactly 3 defectives $= {}^5C_3 \times (0{\cdot}016)^3 \times (0{\cdot}984)^2$

Probability of exactly 4 defectives $= {}^5C_4 \times (0{\cdot}016)^4 \times (0{\cdot}984)$
Probability of exactly 5 defectives $= (0{\cdot}016)^5$

* Depending on the result of the earlier drawings.

Notice that these probabilities are the terms in the Binomial expansion of

$$(0{\cdot}984+0{\cdot}016)^5$$

It is for this reason that these probabilities are said to be Binomially distributed and sampling under such conditions is called Binomial sampling. The Binomial distribution is considered in more detail in Chapter 6.

Definition 3. If the probability of success (p), and hence of failure (q), remains constant throughout a series of n trials, then the probability of obtaining exactly r successes, denoted by $P(r)$, is given by the term containing p^r in the Binomial expansion of $(q+p)^n$, that is

$$P(0) = q^n$$

$$P(1) = {}^nC_1 q^{n-1}p = nq^{n-1}p$$

$$P(2) = {}^nC_2 q^{n-2}p^2 = \frac{n(n-1)}{2!}q^{n-2}p^2$$

etc.

$$P(r) = {}^nC_r q^{n-r}p^r = \frac{n!}{r!(n-r)!}q^{n-r}p^r$$

etc.

$$P(n) = p^n$$

and

$$(q+p)^n = q^n + nq^{n-1}p + \frac{n(n-1)}{2!}q^{n-2}p^2 + \ldots$$

$$= P(0)+P(1)+P(2) \qquad + \ldots$$

In our example $n = 5$; $q = 0{\cdot}984$; $p = 0{\cdot}016$, therefore

$$P(0) = (0{\cdot}984)^5 = 0{\cdot}9225$$
$$P(1) = {}^5C_1(0{\cdot}984)^4(0{\cdot}016) = 0{\cdot}0750$$
$$P(2) = {}^5C_2(0{\cdot}984)^3(0{\cdot}016)^2 = 0{\cdot}0024$$
$$P(3) = {}^5C_3(0{\cdot}984)^2(0{\cdot}016)^3 = 0{\cdot}0001$$

$P(4)$ and $P(5)$ are both negligibly small.

These figures mean that if we took a random sample of 5 from the batch on 10,000 separate occasions, we would expect to find 0 defectives about 9225 times, 1 defective about 750 times, 2 defectives about 24 times and 3 defectives only once. While we would not expect to find a sample with 4 or 5 defectives we realize that it *can* happen. These figures only show what is to be expected, they are not guaranteed to happen.

Cumulative Binomial probabilities are set out, for a range of values of n and p, in Table A.

For example, the probability that a sample of 20 pieces, taken at random from a large batch containing 5 per cent defectives, will contain 2 or less defectives, is found in Table A using $n = 20$, $p = 0{\cdot}05$ and $r = 2$; i.e., the probability is 0·9245.

Again, the probability that another sample of the same number of pieces, from the same batch, will contain exactly 2 defectives is the difference between the probabilities of it containing (a) 2 or less and (b) 1 or less defectives; i.e., the probability is $0{\cdot}9245 - 0{\cdot}7358 = 0{\cdot}1887$.

1.7 Sampling—small batches

If the batch is small, or the sample large, the assumption that the probability of drawing a defective remains constant throughout the sample is no longer valid.

Let us consider again our first box of 1000 set-screw blanks, of which 16 were defective, and examine what happens when we take a random sample of 5 blanks from this smaller batch.

Probability that first blank is defective $= \dfrac{16}{1000} = 0{\cdot}016$

Probability that second blank is defective $= \dfrac{16}{999}$ or $\dfrac{15}{999}$*

$= 0{\cdot}016\,016$ or $0{\cdot}015\,015$

etc.

Probability that the fifth blank is a defective* lies between 16/996 and 12/996, i.e., 0·016 064 and 0·012 048.

We see that this probability is not even approximately constant and our problem is, in such circumstances, somewhat more complex.

Such conditions will not be examined further at this stage; it is sufficient to know that sampling plans—with their associated probabilities—can be determined and such plans are available in a variety of standard sampling inspection tables. However, it must be emphasized again that Binomial sampling only applies to large batches and small samples. It will be appreciated that the terms 'large batch' and 'small sample' are purely relative and no arbitrary numerical limits can be set; a great deal depends on the proportion of defectives thought to exist in the batch.

For example, we have seen that a sample of 5 from a batch of 1000

* As before, this depends on the result of the earlier drawing.

containing 1·6 per cent defectives would *not* conform to the Binomial probability distribution. Consider, in contrast, an unlikely case in which we still take 5 from 1000 but the batch now contains 50 per cent defectives.

$$\text{Probability that first is defective} = \frac{500}{1000} = 0{\cdot}5000$$

$$\text{Probability that second is defective} = \frac{500}{999} \text{ or } \frac{499}{999}$$

$$= 0{\cdot}5005 \text{ or } 0{\cdot}4995$$

etc.

Probability that fifth is defective lies between 500/996 and 496/996, i.e., 0·5020 and 0·4980.

This shows that, although batch and sample size are unchanged, because the proportion defective is much greater we now have approximately constant probability of drawing a defective.

MATHEMATICAL NOTES

1.8 Proof of Addition Law of probability

Law: If E_1 and E_2 are mutually exclusive events with probabilities p_1 and p_2 respectively, then the probability p that either E_1 or E_2 will happen is

$$p = p_1 + p_2 \qquad (1.2)$$

Proof: Consider an action which can produce any one of n equally likely and mutually exclusive results, of which

r_1 produce event E_1
r_2 produce event E_2

$$\text{Therefore, probability of } E_1 \text{ or } E_2 = \frac{r_1 + r_2}{n}$$

$$= \frac{r_1}{n} + \frac{r_2}{n}$$

i.e., $p = p_1 + p_2$.

1.9 Proof of Multiplication Law of probability

Law: If E_1 and E_2 are two independent events with associated probabilities p_1 and p_2, then the probability that both E_1 and E_2 will occur is

$$p = p_1 p_2 \qquad (1.4)$$

Proof: Consider a first trial which has n_1 possible results, of which r_1 produce event E_1; also a second trial which has n_2 possible results, of which r_2 produce event E_2.

There are $n_1 n_2$ possible pairs of results from the two trials, of which $r_1 r_2$ produce the pair E_1 and E_2.

$$\text{Hence } p = \frac{r_1 r_2}{n_1 n_2} = \frac{r_1}{n_1}\frac{r_2}{n_2}$$

i.e., $$p = p_1 p_2$$

1.10 The $\sum$ notation

When a lot of similar terms are to be added together it is much less laborious to write Σ—meaning 'the sum of all terms like the one following'—than to write all the terms.

For example, $x_1 + x_2 + x_3 + \ldots + x_n$ can be written

$$\sum_{i=1}^{i=n} x_i$$

(Σ is the Greek letter *Sigma*).

1.11 The factorial notation

The notation $n!$ is used to denote the product of all the positive whole numbers from 1 up to and including n, i.e., $n!$ = The product of the first n natural numbers.

For example $5! = 5 \times 4 \times 3 \times 2 \times 1 = 120$
$6! = 6 \times 5 \times 4 \times 3 \times 2 \times 1 = 720$

etc.

It is useful to note, and remember, that

$$n! = n(n-1)! \tag{1.6}$$

($n!$ is pronounced n factorial).

1.12 Permutations and combinations

${}^nP_r$ is the symbol used to denote the total possible number of different arrangements (permutations) of n different items, taken r at a time. Differences in sequence of the same set of items constitute different permutations.

${}^nC_r$ is the symbol used to denote the total possible number of different groups (combinations) of n different items, taken r at a time. Different sequences of the same set of items all constitute one and the same combination.

It may help to remember that ordinary usage of the words permutation and combination reverses their meanings, i.e., a lock which depends on an arrangement (permutation) of numbers is called a combination lock, while a football pool which depends on the selection of a group (combination) of matches expected to be draws is called a permutation. Formulae for evaluating ^nP_r and ^nC_r are as follows

$$^nP_r \equiv \frac{n!}{(n-r)!} \tag{1.7}$$

$$^nC_r \equiv \frac{n!}{r!(n-r)!} \tag{1.8}$$

1.13 The Binomial Theorem

The Binomial Theorem states that

$$(a+b)^n \equiv \sum_{r=0}^{r=n} {}^nC_r a^{n-r} b^r \tag{1.9}$$

if n is a positive whole number.

If n is *not* a positive whole number the expansion is similar to eq. (1.9) but the upper limit is $r = \infty$ instead of $r = n$, i.e., an infinite series is produced.

Also, since ^nC_r is physically meaningless when n is not a positive whole number, it should be replaced by $\binom{n}{r}$, where, by definition,

$$\binom{n}{r} \equiv \frac{n(n-1)(n-2)\ldots \text{to } r \text{ terms}}{r!} \tag{1.10}$$

1.14 Worked examples

a. 10 gauges, 7 new and 3 old, are kept in identical boxes on a shelf in stores. The storeman takes two boxes at random from the shelf; determine the probability that they contain (a) 2 old gauges, (b) one old, one new gauge.

(a) 10 boxes, each equally likely to be taken from the shelf, therefore $n_1 = 10$.

Three boxes contain old gauges, therefore $m_1 = 3$. Probability that a box, chosen at random, will contain an old gauge is therefore

$$p_1 = \frac{m_1}{n_1} = \frac{3}{10}.$$

When one box, containing an old gauge, has been taken, we are left with 9 boxes and 2 old gauges, therefore

$$n_2 = 9,\ m_2 = 2$$

and probability that next box, chosen at random, will contain an old gauge is

$$p_2 = \frac{m_2}{n_2} = \frac{2}{9}.$$

Hence, by the Multiplication Law, probability that both boxes will contain old gauges is

$$p_1 \times p_2$$

$$= \frac{3}{10} \times \frac{2}{9} = \frac{1}{15}$$

(b) For the boxes to contain one old, one new gauge, either the first box contains a new gauge and the second an old one, or vice versa. Probability that the first box contains a new gauge

$$= q_1 = 1 - p_1 = 1 - \frac{3}{10} = \frac{7}{10}$$

If the first box contains a new gauge there are still 3 old ones on the shelf and

$$p_2' = \frac{3}{9}$$

Hence, probability that first is new and second old

$$= q_1 \times p_2' = \frac{7}{10} \times \frac{3}{9} = \frac{7}{30}$$

If the first box contained an old gauge, then 7 of the remaining 9 boxes contain new gauges, so probability that the second box contains a new gauge is

$$q_2 = \frac{7}{9}$$

(again, we could say $q_2 = 1 - p_2 = 1 - \frac{2}{9} = \frac{7}{9}$).

Hence, probability that first is old and second new

$$= p_1 \times q_2 = \frac{3}{10} \times \frac{7}{9} = \frac{7}{30}$$

so probability that either box contains a new gauge and the other contains an old one

$$= p_2q_1 + p_1q_2$$

$$= \frac{7}{30} + \frac{7}{30} = \frac{14}{30} = \frac{7}{15}$$

(By the Addition Law.)

b. A particular type of aero-engine may be assumed to have a probability of 0·95 of completing a nine-hour flight without malfunction. If the power of the engine is such that a two-engined aircraft can remain airborne on one engine, and a four-engined aircraft on two, what is the probability that (a) a two-engined aircraft, (b) a four-engined aircraft, will be forced down, by reason of engine failure, on a nine-hour flight.

(a) Two-engined aircraft will only be forced down if *both* engines fail.
Let p = probability that one engine fails

$$= 1 - 0{\cdot}95 = 0{\cdot}05$$

By the Multiplication Law, probability that *both* engines fail

$$= p \times p = 0{\cdot}05 \times 0{\cdot}05$$
$$= 0{\cdot}0025$$

(b) Four-engined aircraft will be forced down if 3 or 4 engines fail.
By Multiplication Law: probability that all 4 engines fail

$$= p^4 = (0{\cdot}05)^4 = 0{\cdot}000\,006\,25$$

probability that 3 engines fail and one *specified* engine does not fail

$$= p^3q = (0{\cdot}05)^3 \times 0{\cdot}95 = 0{\cdot}000\,118\,75$$

where $q = 1 - p = 0{\cdot}95$.

But *any* one of the 4 engines could be the one which fails, so by Addition Law: probability that 3 engines fail and *any* one engine does not fail

$$= 4p^3q = 4 \times 0{\cdot}000\,118\,75$$
$$= 0{\cdot}000\,475$$

Using the Addition Law again we get:
Probability that 3 or 4 engines fail

$$= 0{\cdot}000\,006\,25 + 0{\cdot}000\,475$$
$$= 0{\cdot}000\,481$$

That is, probability that two-engined aircraft is forced down

$= 2{\cdot}5 \times 10^{-3}$ and probability that four-engined aircraft is forced down $= 0{\cdot}5 \times 10^{-3}$. Thus, the two-engined aircraft is five times more likely to be forced down than the four-engined aircraft.

c. A firm requires 6 men for a particular task. They have an available labour pool of 10 men, all equally suitable for the task. In how many different ways could the gang of 6 men be chosen?

Our problem is to determine the number of different ways of choosing 6 from 10, all different. Since the sequence in which the men are selected has no bearing on the problem, we require the number of combinations (not permutations).

Number of different gangs of 6 men which can be chosen from 10

$$= {}^{10}C_6$$

$$= \frac{10!}{6!(10-6)!}$$

$$= \frac{10 \times 9 \times 8 \times \ldots \times 3 \times 2 \times 1}{(6 \times 5 \times \ldots \times 3 \times 2 \times 1) \times (4 \times 3 \times 2 \times 1)}$$

$$= \frac{10 \times 9 \times 8 \times 7}{4 \times 3 \times 2 \times 1}$$

$$= 210$$

Thus there are 210 different ways of forming the gang.

d. Consider now a somewhat different problem. Again 6 men are required, but 2 are to be qualified electricians and 4 are to be carpenters. Of the 10 men available 3 are electricians and the remaining 7 are carpenters. Determine the number of ways of forming the required gang.

Number of ways of choosing 2 electricians from 3

$$= {}^{3}C_2$$

$$= \frac{3!}{2!1!} = 3$$

Number of ways of choosing 4 carpenters from 7

$$= {}^{7}C_4$$

$$= \frac{7!}{4!3!}$$

$$= \frac{7 \times 6 \times 5}{3 \times 2 \times 1} = 35$$

Now any pair of electricians can be teamed with any 4 carpenters, therefore number of different gangs $= 3 \times 35 = 105$. That is, there are 105 different ways of forming the gang.

e. For identification purposes components are to be given a four-letter code. Using all the letters of the alphabet (i.e., 26), how many different codes are available:

(a) If no letter is to be repeated in a code.
(b) If a letter may be used 2, 3, or 4 times in one code.

In this problem the sequence *is* of importance, since ABCD is a different code from ACDB for instance. Consequently we are dealing with permutations (not combinations).

(a) Number of ways of choosing and arranging 4 from 26, all different

$$= {}^{26}P_4$$

$$= \frac{26!}{(26-4)!}$$

$$= \frac{26!}{22!}$$

$$= 26 \times 25 \times 24 \times 23$$

$$= 358{,}800$$

so 358,800 codes of 4 letters, with no repetitions, are available.

(b) If repeated letters are allowed we must think carefully and, having seen that the permutation formula does not apply in this case (since the letters are not 'all different') use our powers of reasoning to solve the problem.

Consider the code as having 4 spaces, each space to be filled by any one of 26 letters. The number of ways of filling the first space is 26. Whatever goes in the first space there are still 26 ways of filling the second space. Hence there are 26×26 different ways of filling the two spaces.

Continuing the argument through three to four spaces we get

Number of ways of filling 4 spaces $= 26 \times 26 \times 26 \times 26$
$= 456{,}976$

456,976 codes of 4 letters, with repetitions, are available.

1.15 Problems

1. A box contains a very large number of bolts. 600 have already been inspected and 12 of these found to be defective. Estimate the probability that

(a) the next bolt inspected will be defective;
(b) the next bolt inspected will not be defective;
(c) only one of the next 2 bolts inspected will be defective;
(d) both of the next 2 bolts inspected will be defective;
(e) neither of the next 2 bolts inspected will be defective.

2. Experience shows that, on the average, 20 per cent of the locating dowels produced by a certain machine will be defective. If 10 dowels are selected at random from a day's output of the machine, calculate the probability that
(a) none will be defective;
(b) exactly 2 will be defective;
(c) more than 2 will be defective;
(d) less than 8 will be defective.
If 1000 samples, each of 10 dowels, were taken, in how many of these samples would we expect there to be
(e) exactly 2 defective dowels;
(f) not more than 2 defective dowels;
(g) all 10 defective dowels.

3. If 8 per cent of the rivets produced by a machine are defective, and random samples of 5 rivets are taken from each of 250 large batches arriving at Inspection, how many of these samples would be expected to contain
(a) 0 defectives;
(b) exactly 2 defectives;
(c) less than 2 defectives.

4. A manufactured article is subject to two inspections, each looking for a different type of fault. The process producing the first feature averages 2 per cent defectives in its output, and the process producing the second feature (quite independently) averages 5 per cent defectives. What percentage of articles would be expected
(a) to fail inspection on one fault only;
(b) to fail inspection on both faults;
(c) to pass both inspections.

5. A batch of 800 washers is thought to contain 6 per cent defectives. The batch is well mixed. Calculate the probability that a random sample of 4 washers will contain
(a) 0 defectives;
(b) exactly 1 defective;
(c) 4 defectives.
Calculate also, for comparison, the approximate probabilities, using the Binomial Distribution.

6. A component is produced in four independent operations, there being no inspection between operations. Operation A produces

8 per cent defectives, operation B—5 per cent, operation C—1 per cent and operation D—3 per cent.

A batch of 100,000 pieces starts operation A. Calculate the number of defect-free pieces you would expect to find after operation D is completed.

Would you expect the number to be any different if an inspection is imposed following each operation and all defective pieces are removed from the batch before the next operation?

7. Four machines, a drill, a lathe, a grinder, and a miller operate independently of each other. Their utilizations are:

Drill	50 per cent	Lathe	40 per cent
Miller	70 per cent	Grinder	80 per cent

(a) What is the probability of both drill and lathe not being used on any interval of time?

(b) What is the chance of all machines being busy?

(c) What is the chance of all machines being idle?

8. A random assembly is made up as follows:

Component X	2 off
Component Y	1 off
Component Z	2 off.

The percentages of defectives in the large batches supplied to the assembler are: X 5 per cent, Y 2 per cent, Z 10 per cent.

(a) What proportion of assemblies will be acceptable?

(b) What proportion of assemblies will be unacceptable?

(c) What proportion of assemblies will contain exactly one defective component?

(d) What proportion of assemblies will contain more than one defective component?

9. A piece of equipment will only function if three components A, B, and C are all working. The probability of A failing during one year is 5 per cent, that of B failing is 15 per cent, and that of C failing is 10 per cent. What is the probability that the equipment will fail before the end of one year?

2. Sampling inspection—theory

2.1 Introduction

Sampling inspection is a method of deciding if a batch of pieces is to be accepted—or rejected—on the basis of the number of defective pieces found in a random sample taken from the batch.

It will be obvious that this method of inspection entails the risk of sometimes making the wrong decision. A batch containing a proportion of defectives small enough to make the batch acceptable may yield a sample containing a high proportion of defectives, hence the batch would be rejected. The reverse is also true.

There is no question of regarding the proportion of defectives found in the sample as a measure of the proportion of defectives present in the batch. The decision to accept or reject the batch is made on the basis of the value of the probability that a batch, containing an acceptably small proportion of defectives, would yield a sample containing the proportion of defectives actually found. Putting it another way, if it is unlikely that a 'good' batch would yield a sample as 'bad' as the random one taken, then the batch is probably 'bad' and therefore to be rejected, and vice-versa.

2.2 Terminology

In order to understand the terminology let us consider a simple single sampling scheme (so called because the accept/reject decision is taken on the basis of a single sample from each batch).

We will consider the batches to be large enough to permit the assumption of Binomial sampling conditions.

a. **Sampling Plan** A typical single sampling scheme would, almost invariably, require a much larger sample than this; however, the basic theory is still the same and so a sample of 10 has been used for the sake of clarity.

'Take a random sample of 10 pieces from the batch and inspect them; if less than 2 are found to be defective, accept the batch, otherwise "reject" the batch.'

Let P = probability that any given batch arriving at Inspection will yield a 'good' sample and hence be accepted.

and p = unknown proportion of defectives in the given batch.

Now:

Probability of obtaining a sample of 10 containing 0 defectives $= P(0)$
$= (1-p)^{10}$.

Probability of obtaining a sample of 10 containing 1 defective $= P(1)$
$= {}^{10}C_1(1-p)^9 p$.

But the batch will be accepted if it contains less than 2 defectives, i.e., 0 or 1, therefore

$$P = P(0)+P(1)$$

i.e.,
$$P = (1-p)^{10} + {}^{10}C_1(1-p)^9 p. \tag{2.1}$$

Thus, P, the probability of acceptance of a batch, is a function only of p, the proportion of defectives in that batch. (p is sometimes referred to as the *batch quality*.) A curve of P plotted against p for any given sampling plan is called the operating characteristic for that plan.

Suppose, since it is seldom, (if ever) possible to achieve perfection,

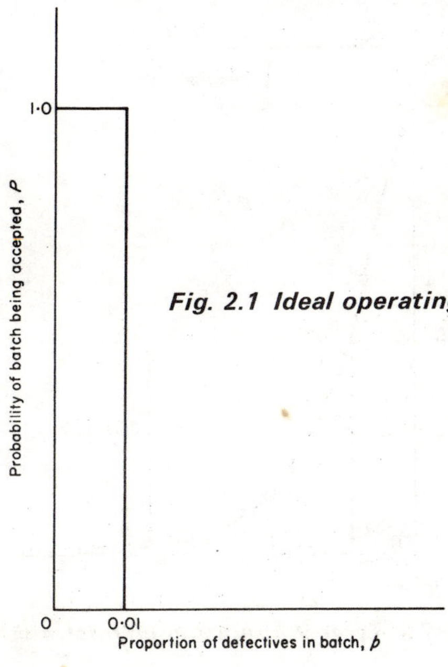

Fig. 2.1 Ideal operating characteristic

that we are prepared to accept batches containing up to, but not exceeding, say 1 per cent of defectives. The ideal sampling plan would have a rectangular operating characteristic, as shown in Fig. 2.1; batches containing not more than 1 per cent of defectives ($p = 0{\cdot}01$) would be certain to be accepted ($P \doteq 1{\cdot}0$); all others would be certain to be rejected ($P = 0$). The only method by which this could be achieved would be by 100 per cent effective inspection of every piece in the batch and this, in practice, is unattainable because of human fallibility. The best we can hope for is a curve which approximates to this shape, as in Fig. 2.2.

b. **Risks** To be of any practical worth our sampling plan must be such that batches containing not more than a specified, small proportion of defectives will be almost certain to be accepted. This small, specified proportion of defectives is called the Acceptable Quality Level (A.Q.L.). The small probability that batches as good as, or better than, the A.Q.L. will yield a bad sample—and hence the batch will be rejected—is called the Producer's Risk (P.R.).

Our sampling plan must also protect the 'consumer'; to this end it is designed to give only a small chance of passing batches which contain a proportion of defectives somewhat larger than the A.Q.L. This larger proportion of defectives is called the Limiting Quality (L.Q.) (also

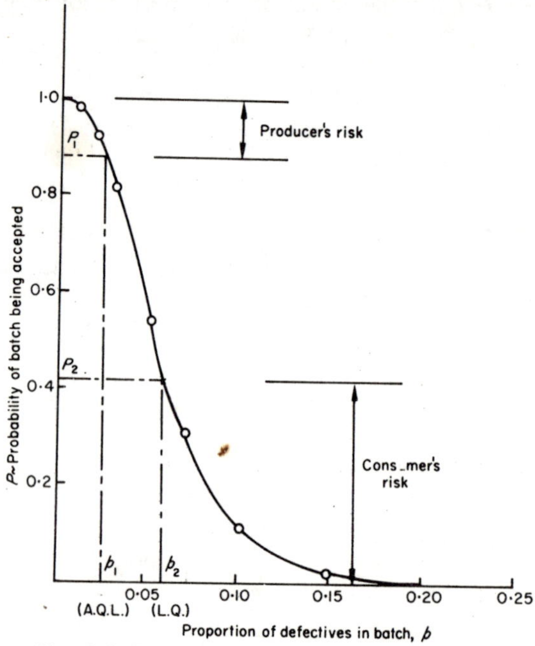

Fig. 2.2 Typical operating characteristic

referred to as the Lot Tolerance Percent Defective—L.T.P.D.). The small probability that a batch as bad as, or worse than, the L.Q. will yield a good sample—and hence the batch will be accepted—is called the Consumer's Risk (C.R.).

In short the Producer's Risk is the probability of rejecting 'good' batches, which ought to have been accepted. The Consumer's Risk is the probability of accepting 'bad' batches, which ought to have been rejected.

All these terms are defined diagrammatically in Fig. 2.2.

2.3 **Choice of sample size**

Still considering our batch to be big enough to permit the assumption of Binomial conditions, the general equation of the operating characteristic is, from eqs. (1.9) and (2.1):

$$P = \sum_{r=0}^{r=a} {}^nC_r q^{n-r} p^r \tag{2.2}$$

where n = number of pieces in the sample;
a = maximum number of defective pieces which may be tolerated in the sample if the batch is to be accepted;
p = proportion of defectives in the batch;
$q = 1-p$;
P = probability of acceptance of a batch containing a proportion p of defectives.

If we specify two pairs of values (P_1, p_1) and (P_2, p_2), then since eq. (2.2) contains 4 unknowns (P, p, n, a), a pair of simultaneous equations can be formed and solved giving values for n and a, that is,

$$1-P_1 = \text{Producer's Risk} \qquad p_1 = \text{A.Q.L.}$$
$$P_2 = \text{Consumer's Risk} \qquad p_2 = \text{L.Q.}$$

Giving

$$\left.\begin{aligned} P_1 &= \sum_{r=0}^{r=a} {}^nC_r(1-p_1)^{n-r}p_1^r \\ P_2 &= \sum_{r=0}^{r=a} {}^nC_r(1-p_2)^{n-r}p_2^r \end{aligned}\right\} \tag{2.3}$$

Solution of these equations is beyond the scope of this book, but it will be seen that, for specified Consumer and Producer's Risks, A.Q.L. and L.Q., our single sampling plan is fully defined and the necessary values of n—the sample size and a—the acceptance number can be determined.

In practice, since n and a must both be whole numbers, it is seldom possible to design a single sampling plan which will achieve exactly the specified values of Producer's Risk/A.Q.L. and Consumer's Risk/L.Q. Some slight departure from the specified values must be expected, and accepted.

It is, of course, equally possible to construct the operating characteristic for a given sample size and acceptance number, hence evaluating the P.R., C.R., A.Q.L., and L.Q. values for that scheme.

2.4 Proportion of batches accepted

By virtue of the empirical definition of probability, the proportion of batches, of any given quality, expected to be accepted by Inspection is numerically equal to the probability of acceptance of a batch of that quality, i.e., P can be interpreted as either

(a) The probability that a single batch, containing a proportion p of defectives, will be accepted

or (b) The proportion of batches, each containing a proportion p of defectives, expected to be accepted by Inspection (taken over a large number of batches).

2.5 Average Outgoing Quality (A.O.Q.)

If 'rejected' batches are 100 per cent inspected and the defectives removed, then the proportion of defectives remaining, taken over all the batches leaving Inspection, will be less than the overall proportion arriving at Inspection—of course!

Average Outgoing Quality (A.O.Q.) is the overall proportion of defectives in a large number of batches leaving Inspection, all the batches having the same proportion, p, of defectives on arrival at Inspection. It assumes that all defectives found have been removed, and the batch size restored by putting in 'good' pieces, from stock.

Let k = number of batches (large);
N = number of pieces in each batch (large);
p = proportion of defectives in each batch;
P = probability of acceptance of each batch or, proportion of batches expected to be accepted (see section 2.4).

Number of batches expected to be 'accepted' $= Pk$
Number of pieces in 'accepted' batches $= NPk$
Number of defectives 'accepted' $= NPkp$
Total number of outgoing pieces $= Nk$
Number of outgoing defectives $= NPkp$

$$\text{A.O.Q.} = \frac{NPkp}{Nk}$$

i.e.,

$$\text{A.O.Q.} = Pp \tag{2.4}$$

Hence A.O.Q. $< p$ as stated earlier.

Our argument assumes that all the defectives in those batches subjected to 100 per cent inspection are found, and extracted. It is, therefore, a little optimistic. On the other hand, no allowance is made for the small number of defectives found in the samples taken from the 'accepted' batches. These defectives would be extracted so, on this score, the argument is a little pessimistic. Both these effects should be small and may be expected to tend to cancel each other out.

If batches arriving at Inspection contain only a small proportion of defectives (p) only very few of them will yield a bad sample and have to be 100 per cent inspected. Consequently the Average Outgoing Quality (p') will be only marginally less than p. When p is somewhat larger, more batches will be 'rejected' and 100 per cent inspected, and p' will be rather less than p. Ultimately, as p gets larger, and the number of batches 100 per cent inspected becomes large, p' will begin to reduce again. Consequently a graph of A.O.Q. against p (see Fig. 2.3) will rise initially, reach a maximum and then A.O.Q. will decrease, as p increases. The maximum value of A.O.Q. is called the Average Outgoing Quality Limit (A.O.Q.L.) and is the worst average quality (over a large number of batches of similar pieces) which may reasonably reach the stores, no matter how bad the quality of the batches when they arrive at Inspection.

This must not be thought to suggest that it is better to produce very bad batches in order to achieve a low value for the average outgoing proportion of defectives! Remember that low A.O.Q. for high p is only

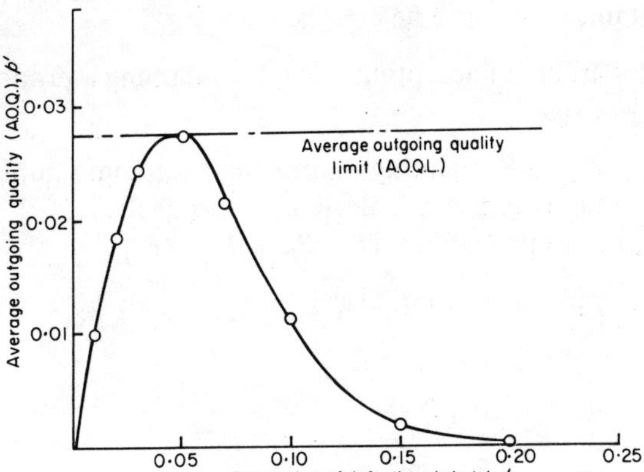

Fig. 2.3 Average outgoing quality curve

achieved at the cost of doing 100 per cent inspection on a large number of batches.

2.6 Non-Binomial sampling

If the conditions necessary for the assumption of the Binomial probability distribution are not satisfied, then the equations for the operating characteristic, etc., will be more complex. However, the general concept remains the same, the necessary calculations have been made, and the appropriate sampling inspection schemes have been published in various forms (D.E.F.–131–A, Dodge-Romig, etc.).

2.7 Worked examples

2.7.1 A single sampling inspection scheme for large batches of mass produced components states:

From each batch take and inspect a random sample of 50 components. If 3 or more are found to be defective, inspect the whole batch and remove all defectives, if less than 3 are found accept the batch without further inspection.

Obtain the equation for P, the probability that a batch, containing a proportion p of defectives, will be accepted, in terms of p

Evaluate P for $p = 0{\cdot}01, 0{\cdot}02, 0{\cdot}03, 0{\cdot}05, 0{\cdot}07, 0{\cdot}10, 0{\cdot}15, 0{\cdot}20, 0{\cdot}25$.

Plot the operating characteristic and the average outgoing quality curve.

Estimate (a) The Producer's Risk for an A.Q.L. of $2\frac{1}{2}$ per cent.
(b) The Consumer's Risk for a L.Q. of 6 per cent.

A batch will be 'accepted' (without further inspection) only if the sample contains 0 or 1 or 2 defectives.

Let P = Probability of accepting a batch containing a proportion p of defectives

$P_{(0)}, P_{(1)}, P_{(2)}$ = Probability of drawing a random sample containing exactly 0, 1, or 2 defectives respectively.

By addition law of probability $P = P_{(0)} + P_{(1)} + P_{(2)}$

Now, for a sample of 50 components

$$P_{(0)} = q^{50} \text{ where } q = 1-p$$

$$P_{(1)} = {}^{50}C_1 pq^{49} = 50pq^{49}$$

$$P_{(2)} = {}^{50}C_2 p^2 q^{48} = \frac{50 \times 49}{2} p^2 q^{48} = 1225\, p^2 q^{48}$$

Therefore

$$P = q^{50} + 50pq^{49} + 1225p^2q^{48}$$
$$= q^{48}(q^2 + 50pq + 1225p^2)$$

But $q = 1 - p$
therefore

$$P = (1-p)^{48}[(1-p)^2 + 50p(1-p) + 1225p^2]$$
$$= (1-p)^{48}[(1-2p+p^2) + 50p - 50p^2 + 1225p^2]$$
$$= (1-p)^{48}\,(1176p^2 + 48p + 1)$$

That is, the equation of operating characteristic is

$$P = (1-p)^{48}\,(1176p^2 + 48p + 1).$$

Now $P = P_{(0)} + P_{(1)} + P_{(2)}$

therefore P is the probability of obtaining 2 *or less* successes (i.e., finding a defective is a 'success') in 50 'trials' (inspecting 50 components).

Hence the values of P for given values of p may be found in Table A, using $n = 50$, $r = 2$.

$p =$	0·01	0·02	0·03	0·05	0·07	0·10	0·15	0·20	0·25
$P =$	0·9862	0·9216	0·8108	0·5405	0·3108	0·1117	0·0142	0·0013	0·0001

We can now plot the operating characteristic (Fig. 2.2) and read off from it when $p = 0{\cdot}025$, $P = 0{\cdot}88$
$p = 0{\cdot}060$, $P = 0{\cdot}42$

i.e., Producer's Risk $= 100 - 88 = 12$ per cent for A.Q.L. $= 2\frac{1}{2}$ per cent and Consumer's Risk $= 42$ per cent for L.Q. $= 6$ per cent.

Now suppose 100 batches arrive at inspection and, unknown to the inspector, each contains 3 per cent of defectives. Since $P = 0{\cdot}8108$ for $p = 0{\cdot}03$, he will find less than 2 defectives in the samples drawn from about 81 of these batches, and pass the batches without further inspection. The remaining 19 batches will probably give samples containing more than 3 defectives, the batches will be 100 per cent inspected and all defectives removed and replaced by 'good' components.

Consequently, the average percentage of defectives getting through is

$$3 \times \frac{81}{100} = 2{\cdot}43 \text{ per cent}$$

This is known as the Average Outgoing Quality (A.O.Q.) i.e.,

$$\text{A.O.Q.} = Pp$$

Tabulating A.O.Q. for all our values of p we have

p	0·01	0·02	0·03	0·05	0·07	0·10	0·15	0·20	0·25
A.O.Q.	0·0099	0·0184	0·0243	0·0273	0·0218	0·0112	0·0021	0·0003	0

We can now plot the average outgoing quality curve (Fig. 2.3).

2.7.2 As Quality Engineer of F Shop you are asked to prepare a single sampling inspection scheme for thread blanks, mass produced in batches of 1500. Since these are not finished products, and will be subject to re-inspection after thread rolling, you decide that an A.Q.L. of $2\frac{1}{2}$ per cent will be suitable. Use DEF–131–A to determine the appropriate sampling plan, and list its characteristics.
Refer to DEF–131–A

Page 8, Para 9.2, 'Unless otherwise specified, Inspection level II will be used.'

Page 11, Batch size 1500 (i.e., 1201 to 3200). Level II. Sample size code letter is K.

Page 12, Sample Size code letter K, therefore sample size = 125. A.Q.L. = $2\frac{1}{2}$ per cent, therefore Acceptance Number = 7.

Thus random samples of 125 blanks must be taken from each batch and inspected. If 7 or less are found to be defective the batch is to be accepted. If 8 or more are found to be defective, the batch is to be rejected.

Page 24, Code letter K. A.Q.L. = $2\frac{1}{2}$ per cent. A.O.Q.L. = 3·6 per cent, that is,

No matter how bad the quality of batches arriving at Inspection, the maximum outgoing proportion defective, averaged over many batches of similar quality, is 3·6 per cent.

Page 28, Code letter K. A.Q.L. = $2\frac{1}{2}$ per cent. L.Q. = 11 per cent for Pa = 5 per cent (Consumer's Risk).

Individual batches arriving at Inspection containing 11 per cent of defectives, or more, have a 5 per cent or less chance of being accepted.

2.8 Problems

1. A sampling inspection scheme states: From each (large) batch take a random sample of 20 pieces; if the sample contains less than 2 defectives accept the batch, otherwise reject the batch.

   Obtain the equation for P, the probability of accepting a batch containing a proportion p of defectives, as a function of p.

Evaluate P for $p = 0{\cdot}01, 0{\cdot}02, 0{\cdot}03, 0{\cdot}05, 0{\cdot}07,$
$0{\cdot}10, 0{\cdot}15, 0{\cdot}20, 0{\cdot}30.$

Plot the operating characteristic.

Estimate (a) The Producer's Risk for an A.Q.L. of 1 per cent.
(b) The Producer's Risk for an A.Q.L. of $6\frac{1}{2}$ per cent.
(c) The Limiting Quality for a Consumer's Risk of 5 per cent.

Plot the corresponding A.O.Q. curve and estimate the A.O.Q.L.

2. For each of the single sampling schemes detailed below determine
(a) The Producer's Risk for an A.Q.L. of $2\frac{1}{2}$ per cent.
(b) The Consumer's Risk for a Limiting Quality of 6 per cent.
Scheme A. Sample size 15. Acceptance number 1.
Scheme B. Sample size 30. Acceptance number 2.

In each case the batches are to be considered large enough to permit the assumption of Binomial sampling conditions.

Which of the two schemes would you consider preferable if you were
(c) The Producer.
(d) The Consumer.

References

D.E.F.–131–A.: *Sampling Procedures and Tables for Inspection by Attributes.* London, H.M.S.O.

Dodge-Romig: *Sampling Inspection Tables.* New York, John Wiley and Sons Inc., 1944.

3. Sampling inspection—variations on a theme

3.1 Single, double, and sequential sampling

Up to now we have considered only single sampling schemes, i.e., schemes which are designed to provide an accept/reject decision on the basis of the number of defective items found in a single sample taken from a batch.

A comparable operating characteristic may be obtained with, on the average, less sampling, by use of a double sampling scheme. Under such a scheme the first sample can lead to one of three decisions

(a) Accept the batch;
(b) Take a second sample;
(c) Reject the batch.

If a second sample is taken, then on the basis of the number of defectives found in the two samples combined, an accept/reject decision is made.

For example, take a first sample of 10 pieces. If 0 defectives are found, accept the batch, if more than 2 defectives are found reject the batch. If 1 or 2 defectives are found in the first sample, take a second sample of 10 pieces, if the *total* number of defectives in both samples is 2 or less, accept the batch; otherwise reject.

Single sample of 10 pieces:

Probability that sample contains 0 defectives $= P(0) = q^{10}$
Probability that sample contains 1 defective $= P(1) = {}^{10}C_1 q^9 p$
Probability that sample contains 2 defectives $= P(2) = {}^{10}C_2 q^8 p^2$.

Now the batch is accepted if:

(a) First sample contains 0 defectives.
or (b) First sample contains 1 defective, second contains 0.
or (c) First sample contains 1 defective, second contains 1.
or (d) First sample contains 2 defectives, second contains 0.

hence the probability that batch is accepted = P where

$$P = P(0)+P(1).P(0)+P(1).P(1)+P(2).P(0)$$
$$= P(0)[1+P(1)+P(2)]+[P(1)]^2$$
$$= q^{10}[1+{}^{10}C_1q^9p+{}^{10}C_2q^8p^2]+[{}^{10}C_1q^9p]^2$$
$$= q^{10}+q^{18}p[{}^{10}C_1q+{}^{10}C_2p+({}^{10}C_1)^2p]$$

i.e., $P = (1-p)^{10}+(1-p)^{18}p[{}^{10}C_1(1-p)+{}^{10}C_2p+({}^{10}C_1)^2p]$ (3.1)

We see once again that P is a function of p only (albeit a more complicated one) and hence an operating characteristic, i.e., a graph of P against p can be drawn.

The decision to accept or reject the batch is made without taking a second sample if the first contains either zero defectives or more than two defectives. Hence, the proportion of batches from which only 1 sample will be taken is

$$P(0)+P(3)+P(4)+P(5)+\ldots+P(29)+P(30 = 1-[P(1)+P(2)]$$

And from the remaining proportion (i.e., $P(1)+P(2)$) of batches 2 samples will be taken.

Consequently, if the first sample is of size n_1 and the second sample is of size n_2, and a large number of batches arrive at Inspection each containing a proportion p of defectives, the *average* number of pieces sampled per batch will be greater than n_1 and less than n_2. This average sample size is denoted by $\bar{n}$ and given by

$$\bar{n} = n_1[1-\{P(1)+P(2)\}]+(n_1+n_2)[P(1)+P(2)]$$
$$= n_1+n_2[P(1)+P(2)]$$

In our example the two sample sizes are equal (i.e., $n_1 = n_2$). This need not necessarily be so and an example using different sample sizes is given in the problems at the end of this chapter.

In general,

$$\bar{n} = n_1+n_2[P(a)+P(a+1)+\ldots+P(b)]$$

where

a = minimum number of 1st sample defectives causing a second sample to be taken.

b = maximum number of 1st sample defectives above which a 'reject' decision is made without a second sample.

(In the preceding example $a = 1$ and $b = 2$.)

Consequently $\bar{n}$ depends on p, since $P(a)$, $P(b)$, etc., depend on p, and hence an average sample size curve, of $\bar{n}$ against p, can be drawn.

In shape it will be very similar to the A.O.Q. curve shown in Fig. 2.3, $\bar{n}$ reaching a maximum at that value of p which makes the first sample most likely to yield between a and b defectives.

A single sampling scheme having the same values for Producer's Risk/A.Q.L. and Consumer's Risk/L.Q. would have a sample size (n) which would be larger than the maximum value of the double sampling scheme average sample size ($\bar{n}$). In addition, since there are two sample sizes and two 'acceptable numbers of defectives' to adjust, a double sampling scheme is more flexible and allows the design of a scheme approximating more closely to the required values of Producer's Risk/A.Q.L. and Consumer's Risk/L.Q.

Still further economy in average sample size can be achieved by the use of a far more sophisticated 'sequential sampling scheme'. We will not involve ourselves in the theory of this scheme at this stage but will be content with an outline of how it would operate in practice.

On a sheet of graph paper a pair of parallel straight lines are drawn. (See Figs. 3.1 a and b). Their positions are determined by reference to sampling inspection tables. (See section 3.5.) Take a single piece at random from the batch. If it is a 'good' one plot one unit vertically, if it is a 'bad' one plot one unit horizontally. So long as the position reached lies in the centre band, sampling continues, the decision to accept or reject the batch only being made when the plotted point moves into either the 'accept' or 'reject' decision region.

Again, the economy of this type of sampling must always be considered in conjunction with its much more involved nature.

3.2 Which scheme to use?

Single sample schemes are the easiest to administer and provide the most readily usable quality information (feed-back, etc.); because of their constant sample size, they are also the most easily understood. They are, however, the least economical in terms of the average number of pieces inspected. This latter consideration may be of paramount importance if the testing is destructive, as for example in the case of explosives, or expensive.

Because, by their very nature, sample size and acceptance number must both be whole numbers it is frequently difficult, or even impossible, to achieve an operating characteristic having the exact parameters called for. Compromise is, therefore, inevitable and published sampling plans are based on either

(a) A.Q.L., for an approximately constant Producer's Risk. Sample size (n) and acceptance number (a) being varied with batch size

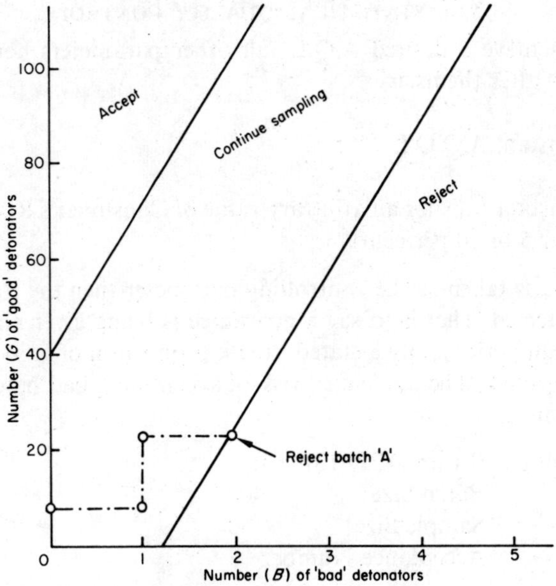

Fig. 3.1. Typical sequential sampling graphs

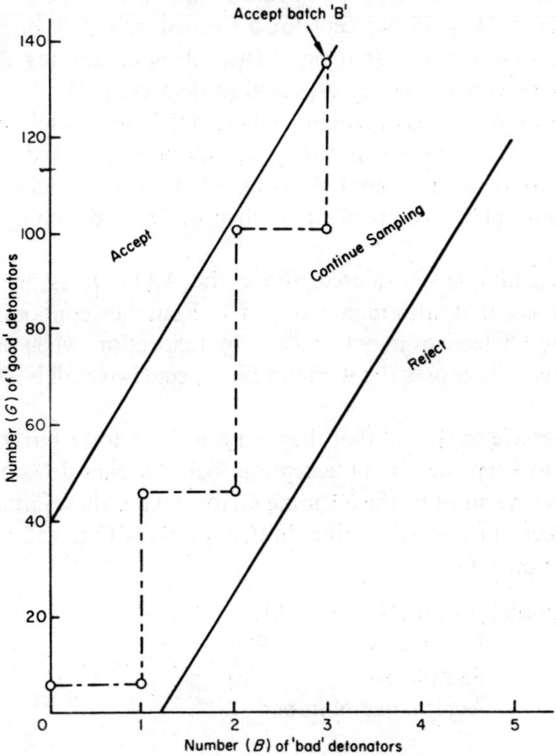

to achieve a desired A.Q.L., all other parameters being left to look after themselves.

or

(b) Constant A.O.Q.L.

or

(c) Constant L.Q. for an arbitrary value of Consumer's Risk (usually either 5 or 10 per cent).

If A.Q.L. is taken as the controlling parameter then the Producer is being protected. That is to say a guarantee is being given that, over a long enough period, only a stated, small, proportion of 'good' batches will be 'rejected'. The associated risk of accepting a bad batch *may* be quite alarming.

For example, (From DEF–131–A):

Batch Size	400
Sample Size	50
Acceptance Number	3

This gives an A.Q.L. = 2·5 per cent with Producer's Risk of 4 per cent, i.e., only 4 batches out of 100 arriving at Inspection containing 2·5 per cent or less of defectives would be expected to be rejected.

However, L.Q. = 15 per cent for a Consumer's Risk of 5 per cent, i.e., 5 batches out of 100 arriving at Inspection containing 15 per cent or more of defectives may be expected to be accepted!

For continuing series of batches, where batch after batch of the same piece will arrive in Stores and they will all be well mixed, the logical parameter to use would seem to be A.O.Q.L. Under these circumstances an odd 'bad' batch loses its identity when well mixed with other 'good' batches.

For the example considered above the A.O.Q.L. is 3·9 per cent. That is to say that, although 5 out of 100 batches containing 15 per cent or more defectives may be passed by Inspection, when mixed with 'good' batches in Stores the worst to be expected overall is 3·9 per cent of defectives.

If batches are so spaced that they are not likely to be mixed, and it is necessary to keep the risk of accepting 'bad' batches down to a small value, then L.Q. must be the deciding factor. Under these circumstances the Producer must, usually, suffer in that a tight A.Q.L. will result from controlling on L.Q.

For example, (From DEF–131–A):

Batch Size	900
Sample Size	80
Acceptance Number	1.

This scheme achieves a L.Q. = 5 per cent for Consumer's Risk of 5 per cent but it imposes an A.Q.L. of 0·4 per cent!

The obvious conclusion to be reached is that a sampling scheme should be chosen which will give a 'quality' as high as necessary, but no higher, since the tighter the limits set the more expensive it will be to achieve them.

3.3 Extensions

It has only been possible here to introduce the topic of sampling inspection, outline the general concept and to define some of the terms used. For more detailed and intricate considerations, such as 'tightened', 'normal', and 'reduced' levels of inspection, classifications of defects as 'critical', 'major', etc., reference should be made to the appropriate sampling inspection documents.

Careful thought should always be given to what is involved; in sampling, for instance, if a batch arrives at the end of a production line to be checked on several dimensions or features, then it may be desirable to fix different quality levels (A.Q.L.?) for different dimensions or features. In contrast, Receiving Inspection have the task of 'over-checking' on inspection already carried out by the Supplier. In this case it may seem reasonable to check at one level only, on the basis of the number of 'defects' missed by the Supplier, separation into dimensional or feature groups being significant only if it suggests a source of error in the Supplier's inspection tackle.

Great care should also be taken to ensure that the sample is as random as reasonably possible and to see that complete 'honesty' is maintained. If a sample can contain 5 defectives and the batch be accepted, while 6 means that the whole batch must be inspected, it would be very tempting to throw back the 6th defective and replace it with a good one! Tempting though it may be the dangers must be only too obvious.

3.4 0/1 Sampling inspection schemes

a. **Introduction** A 0/1 sampling inspection scheme is a single sampling scheme in which the 'acceptable number of defectives in the sample' is always zero (i.e., $a = 0$).

Whereas the usual single sampling schemes achieve various A.Q.L.s by varying a and keeping the sample size n constant, 0/1 schemes keep $a = 0$ and vary A.Q.L. by changing the sample size n.

The advantages of 0/1 schemes lie in their simplicity and psychological appeal—they suggest, quite wrongly of course, that since the samples are free of defects the accepted batches will also be defect-free. They suffer from the overwhelming disadvantage of a badly shaped operating characteristic (see Fig. 3.2). Because of this bad shape, reasonable Producer's Risk/A.Q.L. can only be achieved at the expense of very poor Consumer's Risk/L.Q. and vice-versa. For example: Regardless of batch size, a 1 per cent A.Q.L. with 10 per cent Producer's Risk entails a L.Q. of 20 per cent for a Consumer's Risk of 10 per cent.

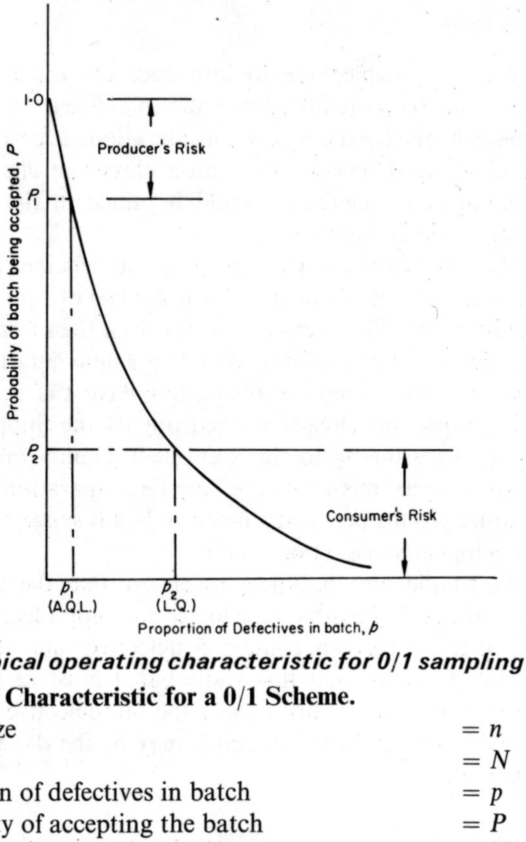

Fig. 3.2 Typical operating characteristic for 0/1 sampling schemes

b. **Operating Characteristic for a 0/1 Scheme.**

Sample size	$= n$
Batch size	$= N$
Proportion of defectives in batch	$= p$
Probability of accepting the batch	$= P$
Number of defectives in batch	$= Np$
Number of 'good' pieces in batch	$= Z = N(1-p)$
Probability that 1st piece is 'good'	$= \dfrac{Z}{N}$
Probability that 2nd piece is 'good', assuming 1st was good	$= \dfrac{Z-1}{N-1}$.
Probability that 3rd piece is 'good', assuming 1st and 2nd were good,	$= \dfrac{Z-2}{N-2}$.

etc.

$$P = \frac{Z}{N}\cdot\frac{Z-1}{N-1}\cdot\frac{Z-2}{N-2}, \text{ etc., to } n \text{ terms} \tag{3.2}$$

If N, and hence also Z, are large enough,

$$\frac{Z}{N} = \frac{Z-1}{N-1} = \frac{Z-2}{N-2} \text{ approx.}$$

$$= \frac{Z-3}{N-3} \text{ approx., etc.}$$

$$= 1-p$$

Now $$\frac{Z}{N} = \frac{N(1-p)}{N}$$

and we have $$P = (1-p)^n \text{ approx.}$$

Equation (3.2) can be expressed more concisely as

$$P = \binom{Z}{n} \div \binom{N}{n} \tag{3.3}$$

3.5 Worked example

You are responsible for the acceptance testing of large batches of detonators. Since the testing is destructive you require a sampling scheme which gives the necessary level of protection against accepting 'bad' batches (i.e., batches containing too many defective detonators) with the minimum amount of testing. Accordingly you decide to use a sequential sampling scheme. Use DEF–131–A to find a suitable scheme for an A.Q.L. of 1 per cent and use it to decide if the following batches A and B are to be accepted or not. Each batch contains 1000 detonators.

Batch A gave the following sequence when the detonators were tested one at a time, at random.

8 good	1 bad	15 good	1 bad

Batch B gave the following sequence

6 good	1 bad	40 good	1 bad
55 good	1 bad	35 good.	

Refer to DEF–131–A.

Page 8 Para 9.2 'Unless otherwise specified Inspection Level II will be used.'
So we will use Level II.

Page 11 Table 1 Batch size = 1000. General Inspection Level II so sample size code letter is J.

Page 66 Table 1–A Code letter J. A.Q.L. = 1·0

H = 40, b = 32

Maximum sample size = 140.*

Page 65 Para 1.2

Score = (Handicap + Number of acceptable units found) − (Penalty × Number of defectives found)

If score reaches a target value equal to twice the handicap, the batch will be accepted. If the score falls to zero or below, the batch will be rejected.

Now, Handicap $H = 40$
so Target $T = 80$
Penalty $b = 32$.

Consider now *Batch A*, and compute the 'score' at each stage

8 good Score = 40 + 8 = 48
Continue testing
1 bad Score = 48 − 32 = 16
Continue testing
15 good Score = 16 + 15 = 31
Continue testing
1 bad Score = 31 − 32 = −1
Score less than zero, reject the batch.

Batch B

6 good Score = 40 + 6 = 46
Continue testing
1 bad Score = 46 − 32 = 14
Continue testing
40 good Score = 14 + 40 = 54
Continue testing
1 bad Score = 54 − 32 = 22
Continue testing
55 good Score = 22 + 55 = 77
Continue testing
1 bad Score = 77 − 32 = 45
Continue testing
35 good Score = 45 + 35 = 80

Score = Target (80), accept the batch without further inspection.

* Maximum sample size = 140. If, after 140 detonators have been tested, a decision has not been reached, then DEF–131–A advises that sequential sampling should be abandoned and a decision made using a multiple sampling scheme. (See DEF 131–A, page 65, para. 2.3.)

N.B. It is worth noting that the characteristics of this scheme are such that, if the first two detonators inspected are defective, the batch is rejected. On the other hand, the first 40 would have to be 'good' before the batch could be accepted.

This same scheme could be operated graphically, as follows:

We have $S = H + G - b \times B$

where S = score,

G = number of acceptable detonators found

B = number of defective detonators found

i.e., $S = 40 + G - 32B$.

We accept the batch if $S \geqslant 80$
and reject the batch if $S \leqslant 0$.

Consider the two equations

$$80 = 40 + G - 32B$$
$$0 = 40 + G - 32B$$

i.e.,

$$G = 32B + 40$$
$$G = 32B - 40$$

These two equations represent parallel lines on a graph of G plotted against B. (See Figs. 3.1, a and b.) Starting from the origin we move one unit to the right for each defective detonator found and one unit upwards for each good detonator tested. If the point reached remains between the lines we continue testing, if it falls on or above the upper line we accept the batch, if it falls on or below the lower line we reject the batch.

In practice each piece would be plotted as 'good' or 'bad' as tested —for clarity Figs. 3.1, a and b show only the points where a change from a run of 'good' detonators was interrupted by a 'bad' one. These figures show that the same decisions are reached graphically as were reached by computing a score.

3.6 Problems

1. A double sampling scheme is as follows. 'Take a first sample of 20; if 0 or 1 defectives are found, accept the batch; if 4 or more defectives are found, reject the batch; if 2 or 3 defectives are found take a further sample of 20 and if, in the combined sample of 40, less than 5 defectives are found, accept the batch; otherwise reject the batch.'

   Write down the equation of the operating characteristic and plot the curve for values of p of

   $$0{\cdot}02,\ 0{\cdot}03,\ 0{\cdot}05,\ 0{\cdot}07,\ 0{\cdot}10,\ 0{\cdot}15,\ 0{\cdot}20,\ 0{\cdot}25.$$

Estimate the proportion of samples expected to contain 2 or 3 defectives (in terms of p) and hence construct a graph of average sample size $\bar{n}$ against p.

Read off (a) the Producer's Risk for an A.Q.L. of 5 per cent;
(b) the Limiting Quality for a Consumer's Risk of 10 per cent.

In each case state also the corresponding average sample size.

2. A double sampling inspection scheme states 'Take a sample of 25 pieces from the batch, if the sample contains 0 or 1 defectives accept the batch, if it contains more than 3 defectives reject the batch. If the first sample contains 2 or 3 defectives take a second sample, of 30 pieces, and accept the batch if the total number of defectives found in both samples does not exceed 3, otherwise reject the batch.' Write down the equation of the operating characteristic for this scheme and use Table A to determine
(a) The Producer's Risk for an A.Q.L. of 2 per cent.
(b) The Consumer's Risk for a L.Q. of 6 per cent.
(c) The A.O.Q. and average sample size for a series of batches each containing 6 per cent of defectives.

3. The boundary lines for a sequential sampling scheme are

$$G = 9+4B \text{ (accept)}$$
$$G = -16+4B \text{ (reject)}.$$

The following sequences were obtained from two batches; use the above sequential scheme to decide, for each batch, if it is acceptable or not.

Batch I GGGG B GGG GGG GG B GGGG B
GGG GGG B G.

Batch II GGG B G BB GGG B GG B G
B GGG GG B GG BB GG BBB
G B.

4. *Barnard's Handicap Method.* This is an alternative method of Sequential Sampling and employs two parameters known as the 'Handicap' and the 'Penalty'. After the inspection of each piece the 'Score' is computed.

Scores (S) = Handicap (H) + Number of 'good' pieces (G)
− [Penalty (P) × Number of 'bad' pieces (B)].
i.e., $S = H+G-PB$

A typical plan might have

Handicap H = 175
Penalty P = 95
Target value = 350.

When the score reaches the target value the batch is accepted, if it falls to zero or below the batch is rejected. A 'safety valve' is included in the form of a maximum sample size (in this instance 675). If this number of pieces is reached (i.e., $G+B = 675$) without the target value, or zero, being reached, then the batch would be rejected.

Compute the 'score' for each of the following sequences and state if the batch is accepted or rejected:

(a) G, B, $16G$, B, $67G$, $2B$.

(b) $107G$, $2B$, $180G$, B, $241G$.

'Translate' the scheme of problem 3 into the 'Handicap' form, determining the values of the Handicap, Penalty and Target. Compute the scores for the sequences given in problem 3, hence checking your previous decisions.

5. Batches of goods, each containing 800 pieces, are inspected according to the following scheme: 'Take a random sample of 10 pieces, if 0 defectives are found in the sample, accept the batch, otherwise reject the batch'.

Calculate the probability that a batch will be accepted if it contains

(a) 1 per cent of defectives;

(b) 15 per cent of defectives.

Determine the corresponding probabilities for batches of 100.

6. From each (large) batch of a certain product, 10 pieces are taken at random by Inspector A. If he finds 0 or 1 defectives he passes the batch.

Inspector B takes an independent random sample of 10 pieces and follows the same procedure.

If the two inspectors agree the batch is accepted or rejected as the case may be. If they disagree a further random sample of 20 pieces is taken and the batch accepted if this latter sample contains not more than 2 defectives.

Obtain, in terms of p, expressions for

(a) The probability that a batch is accepted.

(b) The average sample size.

Determine

(c) The Producer's Risk for an A.Q.L. of 5 per cent.

(d) The Limiting Quality for a Consumer's Risk of 10 per cent.

State also the average sample size for each case.

4. Frequency distributions

4.1 Introduction

If we have a large quantity of related data, for example the diameters of 5000 ball bearings all nominally the same size, or the weights of 200 castings from similar moulds, each diameter or weight is liable to occur more than once. The number of times that a diameter (or weight) occurs is known as the 'frequency' of that size and a table of diameters, with their appropriate frequencies, arranged in ascending diameters is called a 'frequency distribution', since it shows how the various ball bearings are distributed throughout the range of diameters.

4.2 Types of data

Data may be classified either as

(a) Discrete, i.e., data which can take only certain values, and nothing in between; or

(b) Continuous, i.e., data which can take any value within specified limits.

For example; a sample of 10 bolts taken from a batch may contain 0, or 1, or 2, etc., defectives; there cannot be $1\frac{1}{2}$ defectives! Thus the number of defective pieces found in a sample is an example of discrete data. Again, these bolts may have a head thickness which varies from 0·196 in. to 0·205 in., any individual bolt head having any thickness between these values. Thus head thickness is an example of continuous data.

A useful rule of thumb would be to say that data obtained by counting are discrete, while data obtained by measurement are continuous.

Because of limitations of measurement accuracy, continuous data often have the appearance of being discrete. The bolts previously mentioned may have their head thicknesses measured to the nearest 0·001 in., hence all bolts would be recorded as having head thicknesses of 0·196, 0·197, 0·198, 0·199 … 0·204, or 0·205 in.

4.3 Grouping of continuous data

It is frequently more convenient, or even unavoidable, to arrange continuous data in groups. For instance, measuring to the nearest 0·001 in. and quoting a head thickness as 0·198 in., we have, in fact, included in one 'group' all those bolts whose head thicknesses fall between 0·1975 and 0·1985 in. How we deal with a value of exactly 0·1975 or 0·1985 in. will depend on which of two conventions we decide to observe. The most usual is to 'round-up' any number ending in 5, in which case 0·1975 would be recorded as 0·198 and 0·1985 as 0·199. The alternative is to use the 'even-integer' convention, rounding any number ending in 5 to the nearest preceding even integer, then 0·1975 would become 0·198 and 0·1985 would also become 0·198. The latter convention is statistically more desirable since it is unbiased, i.e., it is reasonable to suppose that—in a large mass of data—as many numbers ending in 5 will be rounded 'down' as are rounded 'up', whereas the former convention rounds 'up' every time. Each group is known as a class, the upper and lower values are known as class boundaries, the 'average' of these boundaries is called the class mid-mark, and the difference between successive mid-marks is called the class interval. The number of pieces falling into a given class is called the frequency in that class.

As an example: twenty round bars of steel (nominal $\frac{1}{2}$ in. diameter) were measured for diameter to the nearest 0·001 in. using the even integer convention. Form the frequency distribution of this data.

TABLE 4.1 All dimensions in inches (to the nearest 0·001 in.)

Bar diameters:

0·497	0·499	0·500	0·497	0·498
0·505	0·502	0·504	0·502	0·500
0·498	0·496	0·501	0·503	0·499
0·500	0·502	0·501	0·497	0·505

TABLE 4.2 Class interval 0·001 in.

Class Boundaries	Class Mid-Mark	Frequency
0·4955		
	0·496	1
0·4965		
	0·497	3
0·4975		
	0·498	2
0·4985		
	0·499	2
0·4995		
	0·500	3
0·5005		
	0·501	2
0·5015		
	0·502	3
0·5025		
	0·503	1
0·5035		
	0·504	1
0·5045		
	0·505	2
0·5055		

Choice of the number of classes, and hence the width of the class interval, is largely a matter of common sense. The number of classes should be large enough to show the shape of the distribution but not so large that the class frequencies become very small and introduce spurious apparent variations which would not be evident had the total frequency been greater. This will be considered in more detail later but, for now, a useful guide is to divide data into between 8 and 20 classes, according to the numerical convenience of class boundaries.

Care must be exercised when determining the class boundary values. For example, a number of pieces of wood, about 6 in. long, are measured with an ordinary ruler and recorded as 5·9, 5·8, 6·1, 6·0 in., etc.

If these are stated to be to the nearest 0·1 in. then the class boundaries are 5·75, 5·85, 5·95 in., etc.

However, an alternative convention is to state 'completed' tenths only, i.e., fractions of tenths are ignored. Then the class boundaries are 5·8 to $5{\cdot}9^-$, 5·9 to $6{\cdot}0^-$, 6·0 to $6{\cdot}1^-$, etc., where the 'minus' sign indicates 'up to but excluding'.

Using this convention, a rod whose length was 5·899 in. would be included in the group 5·8 to $5{\cdot}9^-$ in., but if the length was 5·900 in. it would be in the group 5·9 to $6{\cdot}0^-$ in.

Thus the method of collection of the data determines the class boundaries.

NB. When grouping discrete data another term must be introduced, namely 'class limits'. These are the highest and lowest discrete values included in each class, e.g., if the discrete values 0, 1, 2, etc., up to 99 were grouped into 20 groups of equal width, the class limits would be 0–4, 5–9, 10–14, etc., while the class boundaries (by analogy with continuous data) would be −0·5 to +4·5, 4·5 to 9·5, 9·5 to 14·5, etc., with class mid-marks 2, 7, 12, etc.

4.4 Pictorial representation

The object of tabulating and grouping data is to make it easier to see the overall pattern of the variation involved. This can often be further helped by representing the data pictorially. This visual presentation can take various forms known as histograms, ogives, etc. and these will be discussed individually in the following paragraphs.

a. **Histograms** A histogram is a column diagram in which the column areas are arranged to be proportional to the frequency of occurrence of the data value represented by that column.

We can best discuss the construction and use of histograms by using a numerical example:

1000 rods, nominal length 8 in., were measured to the nearest 0·1 in. using the even integer convention (see page 41) and the results tabulated, as follows

TABLE 4.3

Class Boundaries	Class Mid-Mark	Frequency
7·75		
	7·8	78
7·85		
	7·9	164
7·95		
	8·0	547
8·05		
	8·1	157
8·15		
	8·25	54
8·35		

All dimensions in inches

Notice that the last class interval is twice the width of the others, consequently the *height* of the histogram column for this interval is only half what it would be for a single width class, having the same frequency, i.e., the area of the column is proportional to the frequency. This amounts to saying that we have assumed the 54 to be divided equally into 27 in the interval 8·15 to 8·25 and 27 in the interval 8·25 to

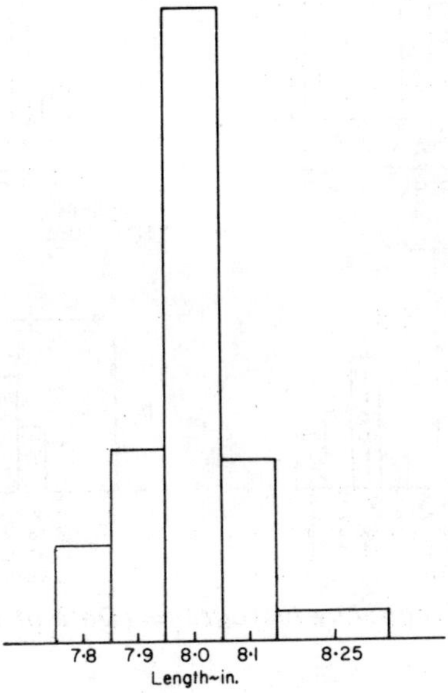

Fig. 4.1 Frequency histogram (Data of table 4.3)

8·35. To indicate that we do not *know* that this subdivision is correct we omit the vertical line dividing this column into two, each of width equal to the other columns.

Consider now another set of values. This time we have the diameters of 800 shafts, measured to 0·0001 in. and grouped in 0·001 in. intervals, using the even integer convention (see page 41), as follows.

TABLE 4.4

Class Boundaries	Class Mid-Mark	Frequency
0·6725		
	0·673	38
0·6735		
	0·674	53
0·6745		
	0·675	98
0·6755		
	0·676	203
0·6765		
	0·677	188
0·6775		
	0·678	111
0·6785		
	0·679	64
0·6795		
	0·680	45
0·6805		

All dimensions in inches

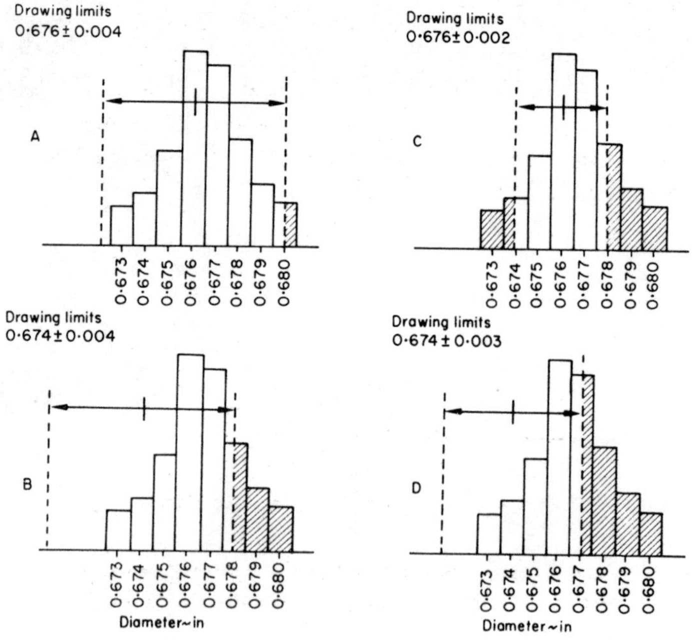

Fig. 4.2 Frequency histograms (Data of table 4.4)

If the specification calls for a diameter of 0·676 in. ±0·004 the histogram shows that the setting is satisfactory and that the process is capable of meeting the tolerances with only a small proportion defective. (The last interval contains diameters between 0·6795 and 0·6805; those in excess of 0·6800 are defective.)

However, if the specification calls for 0·674 ± 0·004 in. we see that the process is capable of meeting the tolerances but that the centre of the distribution is incorrectly set, something like 0·002 in. too little being taken off. Again, had the specification called for 0·676 ± 0·002 in. we would have seen that, although the process was correctly set at mid-tolerance, it was too variable to meet the required limits consistently.

There is yet another situation which can arise; had the specification called for 0·674 ± 0·003 in. we would have seen that a considerable proportion of defectives was inevitable but that the proportion could be minimized by adjusting the setting to bring the centre of the distribution down to 0·674 in., i.e., mid-tolerance.

Histograms may be drawn either for actual frequencies or—by dividing each frequency by the total—for relative frequency (i.e., probability). If two histograms are to be compared with each other then relative frequency must be used, otherwise, different total frequencies may give a misleading impression of differences between the distributions, which do not really exist.

When data are discrete (e.g., number of defectives found in samples of a given size, taken from a batch of components) it is common practice to use a column of unit width for each discrete value, marking that discrete value at the centre of the width.

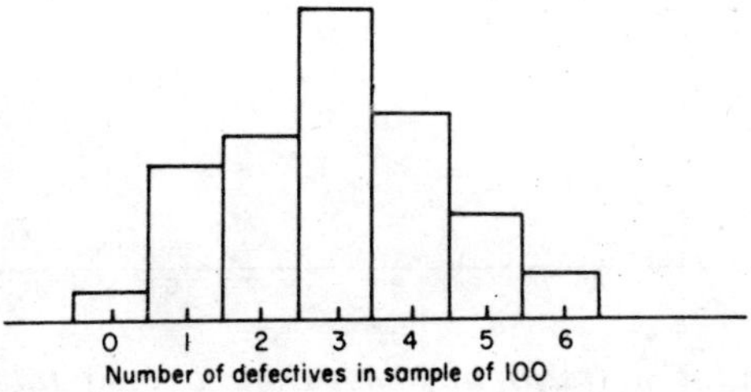

Fig. 4.3 Histogram for discrete data

It must be clearly understood that this does *not* imply grouped data, i.e., the '3 defectives' column does *not* refer to values between 2·5 and 3·5.

b. **Cumulative Frequency Ogive** Consider again the 1000 rods listed in Table 4.3.

78 have a length less than 7·85 in.
78 + 164 = 242 have a length less than 7·95 in., etc.

These are called 'cumulative frequencies' *less than* the class upper boundary and are plotted as a curve, known as the cumulative frequency (less than) ogive.

TABLE 4.5

Class Upper Boundary	Cumulative Frequency Less Than Class Upper Boundary
7·85	78
7·95	242
8·05	789
8·15	946
8·35	1000

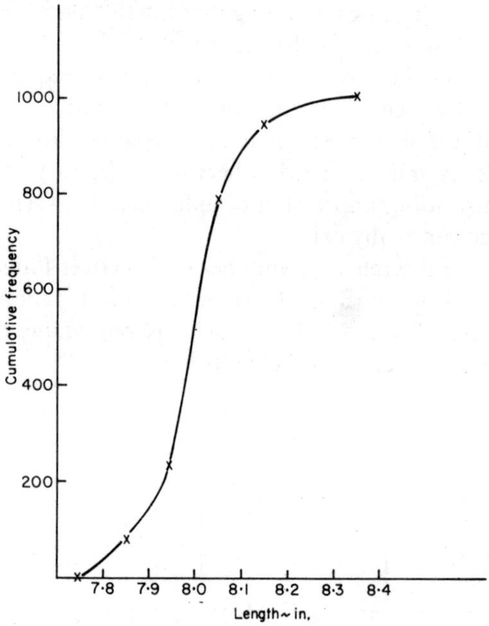

Fig. 4.4 Cumulative frequency 'less than' ogive (Data of table 4.5)

It is important to note that Fig. 4.4 is a 'less than' ogive, i.e., it contains all the values in a given class since they are all less than the upper boundary of the class. For this reason the cumulative frequency must be plotted against the class upper boundary and not against the class mid-mark. For some purposes it is useful to plot a 'greater than' ogive, in which case the cumulative frequency will diminish instead of increasing and must be plotted against class lower boundary. The 'greater than' ogive for the same data is shown in Fig. 4.5.

TABLE 4.6

Class *Lower* Boundary	Cumulative Frequency Greater Than Class Lower Boundary
7·75	1000
7·85	922
7·95	758
8·05	211
8·15	54
8·35	0

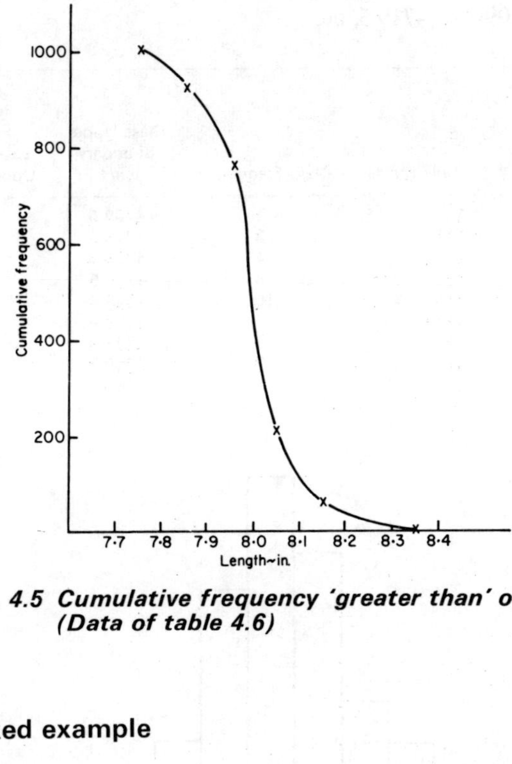

Fig. 4.5 Cumulative frequency 'greater than' ogive (Data of table 4.6)

4.5 **Worked example**

A machine is used to fill bottles which are supposed to contain 4·75 litres (4750 cm^3). A random sample of 40 bottles was taken and the contents of each measured to the nearest cm^3. Prepare grouped frequency distribution and cumulative percentage frequency tables for these data and hence a histogram and cumulative frequency (less than) ogive.

Volume of liquid -cm³.

4731	4746	4755	4720	4745
4749	4762	4786	4744	4739
4738	4757	4763	4724	4776
4714	4773	4712	4737	4748
4755	4756	4718	4700	4783
4757	4724	4732	4739	4742
4735	4763	4741	4740	4730
4748	4749	4727	4751	4751

Greatest volume = 4786 cm^3
Least volume = 4700 cm^3

It will be adequate, and convenient, to use 9 groups, 4700 to 4709 inclusive, 4710 to 4719 inclusive, etc., Since the data have been obtained by measuring to the nearest cm^3, the real class boundaries are 4699·5 to 4709·5, 4709·5 to 4719·5, etc.

Class Limits cm^3	Tally Marks	Class Frequency	Class Upper Boundary cm^3	Cumulative Percentage Frequency Less than class Upper boundary
4700–4709	1	1	4709·5	2·5
4710–4719	111	3	4719·5	10
4720–4729	1111	4	4729·5	20
4730–4739	~~1111~~ 111	8	4739·5	40
4740–4749	~~1111~~ ~~1111~~	10	4749·5	65
4750–4759	~~1111~~ 11	7	4759·5	82·5
4760–4769	111	3	4769·5	90
4770–4779	11	2	4779·5	95
4780–4789	11	2	4789·5	100

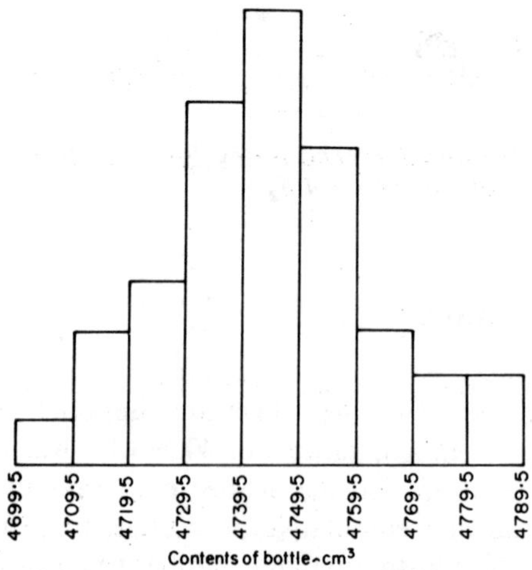

Fig. 4.6 Frequency histogram (Data of section 4.5)

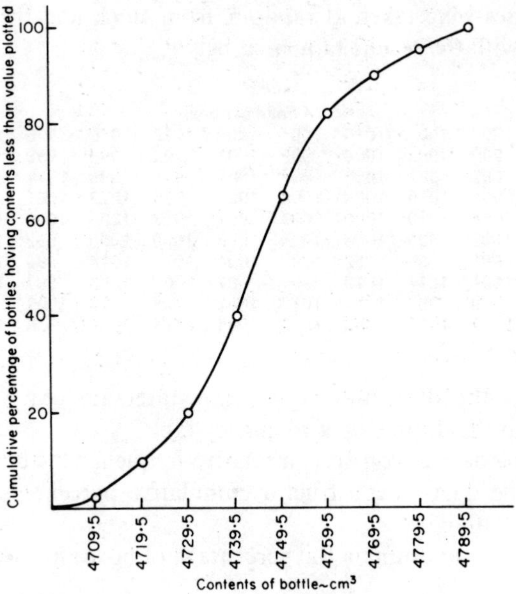

Fig. 4.7 Cumulative percentage frequency (less than) ogive (Data of section 4.5)

4.6 Problems

1. The table shows the distribution of breaking loads in tonf for cables of a given diameter produced by a certain company.

Breaking Load tonf	No. of cables
9·3– 9·7	2
9·8–10·2	5
10·3–10·7	11
10·8–11·2	18
11·3–11·7	14
11·8–12·2	6
12·3–12·7	3
12·8–13·2	1
	60

Illustrate this data graphically by means of
(a) a histogram;
(b) a cumulative frequency ogive.

2. Bolts are supplied by a certain manufacturer in boxes containing a *nominal* 1000. In fact, the bolts are not counted when packed, the system used is to simply fill the box—consequently the actual number varies from box to box.

100 boxes were taken at random, from stock and their contents counted, with the results tabulated below.

No. of bolts per box:

992	1007	1001	1010	1001	1008	1013	1010	995	1003
1009	990	1003	1008	999	1002	992	990	998	1013
1014	992	991	1006	998	1001	991	998	1011	985
986	996	1010	1008	1008	1024	995	1023	1001	998
998	1004	1010	1000	1002	997	1013	1005	1006	1015
994	1002	999	1008	1025	998	1005	1008	992	1016
1003	981	1005	982	997	1004	1013	1018	984	1004
1009	1001	1014	1018	991	997	1009	1003	998	1007
1028	1000	988	1012	1012	1001	979	992	1002	989
1010	1017	1011	996	1006	1009	994	1011	976	1003

(a) Group the data into equal class intervals—e.g., 1000–1004 inclusive and draw up a frequency table.
(b) Use the data to construct a relative frequency histogram.
(c) Use the data to construct a cumulative percentage frequency (less than) ogive.

Use the ogive to estimate the percentage of boxes in stock likely to contain
(i) less than 1000 bolts,
(ii) more than 1010 bolts.

3. An assembled product consists of two main components A and B having resistances R_1 and R_2 the resistance of the assembly then being R_1+R_2. Random samples of the components, each consisting of 30 items, were drawn from stock and their resistances (in ohms) measured.

Component A:	3·14	3·17	3·21	3·15	3·22	3·13	3·15
	3·09	3·18	3·19	3·16	3·24	3·11	3·22
Resistance	3·17	3·06	3·23	3·14	3·12	3·27	3·19
(R_1)	3·17	3·14	3·20	3·19	3·13	3·25	3·16
	3·11	3·17					
Component B:	2·09	2·05	2·10	2·02	2·06	2·13	2·11
	2·07	2·08	1·99	2·07	2·12	2·10	2·03
Resistance	2·05	2·08	2·08	2·00	2·07	2·05	2·13
(R_2)	2·09	2·05	2·07	2·04	2·15	2·09	2·05
	2·13	2·08					

Group the data to form frequency distributions and hence draw histograms, for the resistances of the two components A and B.

A and B have specified resistances of $3{\cdot}1 \pm 0{\cdot}15$ ohms and $2{\cdot}05 \pm 0{\cdot}07$ ohms respectively. What conclusions do you draw from the frequency distributions of the sample resistances?

4. The following table shows the diameters (in inches) of a sample of 60 ball bearings manufactured by a certain process. Construct a

frequency distribution of the diameters using class intervals of 0·005 in. (0·720–0·724 etc.). Also construct a histogram, and a percentage ogive.

0·738	0·729	0·743	0·740	0·736	0·741	0·735	0·731	0·726	0·737
0·728	0·737	0·736	0·735	0·724	0·733	0·742	0·736	0·739	0·735
0·745	0·736	0·742	0·740	0·728	0·738	0·725	0·733	0·734	0·732
0·733	0·730	0·732	0·730	0·739	0·734	0·738	0·739	0·727	0·735
0·735	0·732	0·735	0·727	0·734	0·732	0·736	0·741	0·736	0·744
0·732	0·737	0·731	0·746	0·735	0·735	0·729	0·734	0·730	0·740

(a) What percentage of the ball bearings produced by this process would you expect to have a diameter
  (i) exceeding 0·732 in.
  (ii) not more than 0·736 in.
  (iii) between 0·730 and 0·738 in.

(b) Would you expect the process to be capable of producing ball bearings to the following specifications?
  (iv) $0·73 \pm 0·02$ in.
  (v) $0·73 \pm 0·015$ in.
  (vi) $0·74 \pm 0·01$ in.

5. Measures of location and scatter

5.1 Introduction

In order to describe a distribution numerically, or to compare two or more distributions, it is necessary to have measures of (a) 'location'—an indication of the central value of the distribution, and (b) 'scatter'—an indication of the variability of the members of the distribution.

The most common—and most useful—measures of location and scatter are known as 'mean' and 'standard deviation' respectively, and will be defined later in the chapter. Suppose two machines are producing studs to the same nominal diameter and samples taken from the output of the two machines give the following data:

Machine	A	B
Mean diameter, in.	0·749	0·751
Standard deviation of diameter, in.	0·005	0·002

This shows that the diameters of the studs produced by machine A vary much more—from piece to piece—than do those from machine B, but that the average diameter of the studs from machine A is 0·002 in. smaller than from machine B.

We will discuss the importance of standard deviation in more detail later but, for the present, it is sufficient to say that—in many instances—the limits of dimensional variability, due to random causes, from piece to piece produced by a mass-production process are mean ± 3 times the standard deviation. That is to say, that the greatest and least dimensions likely to be produced are 3 standard deviations greater than, and 3 standard deviations less than, the mean respectively. It must be stressed that we are talking about *random* variation from piece to piece—not progressive change due to an assignable cause such as tool wear.

5.2 Notation

To distinguish between population parameters and sample parameters the symbols μ and σ will be used to denote the population mean and standard deviation respectively, while $\bar{x}$ and s will be used to denote the mean and standard deviation of a sample taken from a population.

The relationships between μ and $\bar{x}$, and between σ and s, will be dealt with more fully in chapter 8. For the moment it will be sufficient to accept that, if $\bar{x}$ and s have been calculated for a large sample (say 40 or more) then they are good estimates of μ and σ.

5.3 Measures of location

a. **Arithmetic Mean** ($\bar{x}$) This is simply the ordinary 'average' of all the values of the variate in the sample.

If x_1 occurs f_1 times, x_2 occurs f_2 times, etc., and lastly x_n occurs f_n times, there being n possible *different* values of x and N observations altogether, i.e.,

$$N = f_1 + f_2 + f_3 + \ldots + f_n \qquad (5.1)$$

then

$$\bar{x} = \frac{f_1 x_1 + f_2 x_2 + f_3 x_3 + \ldots + f_n x_n}{f_1 + f_2 + f_3 \ldots + f_n} \qquad (5.2)$$

This is written, using the Σ notation, as

$$\bar{x} = \frac{\Sigma fx}{\Sigma f} \qquad (5.3)$$

or

$$\bar{x} = \frac{1}{N}\Sigma fx \quad \text{since } N = \Sigma f \qquad (5.4)$$

b. **Median** The median is that value of the variate which divides the total frequency exactly in half, i.e., half the values of the variate are greater, and half less, than the median value. (See Fig. 5.1.)

If there is an odd number $(2N+1)$ of observations of the variate, and they are arranged in ascending order, then the median is the $(N+1)$th value.

If the number of observations is even $(2N)$ then the median is—by convention—taken to be the average of the two middle values, when arranged in ascending order, that is,

$$\text{median} = \frac{N\text{th} + (N+1)\text{th value}}{2}$$

This may perhaps be remembered as 'the mean of the mid-most'.

c. **Mode** The mode is that value of the variate which occurs with the greatest frequency. (See Fig. 5.1.)

d. **Relationship Between Mean, Median, and Mode** For a symmetrical distribution the mean, median, and mode will all be the same value of the variate.

It can be shown that, when a distribution is only moderately skewed, there is an approximate relationship between the three measures of location which may be expressed as:

$$(\text{Mean} - \text{Mode}) = 3(\text{Mean} - \text{Median}) \tag{5.5}$$

(See Fig. 5.1.)

e. **Quartiles, Deciles, etc.** Quartiles are those 3 values of the variate which divide the whole distribution into 4 equal parts. Brief reflection will show that the middle quartile is the median.

Deciles are those 9 values of the variate which divide the whole distribution into 10 equal parts.

Definitions of the terms Octiles, Percentiles, etc., follow along similar lines and will be left unsung.

5.4 Measures of dispersion (or scatter)

a. **Range and Interquartile Range** An obvious, and simple, measure of variability within a sample is the overall range of the observed values. That is to say that sample range is the numerical difference between the greatest and least values in the sample.

Sample range (w) = Greatest value of variate found in sample − least value found.

However, a single, very high (or very low), value of the variate—quite out of character with the remainder of the observations—could have a vast effect on the range of the sample. Consequently 'range' is a very inefficient measure of variability and interquartile range, the difference between the upper and lower quartile values, is used in preference, since it defines the width of the band within which the central 50 per cent of the observations lie, and is almost unaffected by one or two stray uncharacteristic values of the variate.

Interquartile range = Upper quartile − lower quartile. (See Fig. 5.1.)

Some authorities prefer to quote half this value, and call it the semi-interquartile range. Unless the distribution is symmetrical this will *not* be equal to the difference between the median and either the upper or lower quartile.

Semi-interquartile range = $\frac{1}{2}$ (Upper quartile − lower quartile).

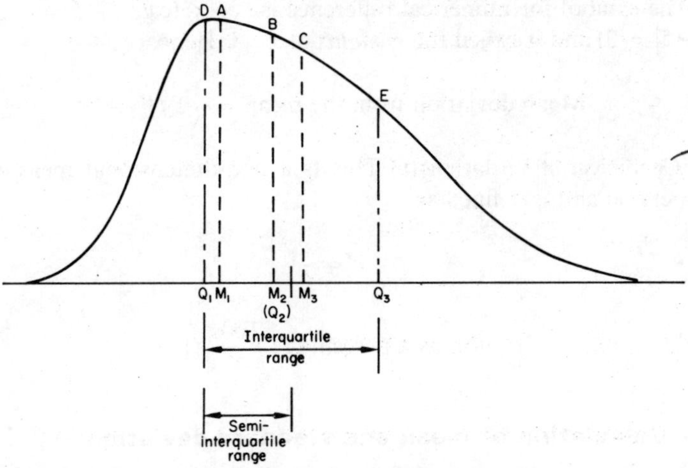

M_1 is the mode (A is the 'peak' of the distribution curve)
M_2 is the median (area to left of BM_2 equals area to the right)
M_3 is the mean (turning moment, about CM_3, of area to the left is equal to the moment of the area to the right)
Q_3 is the upper quartile
Q_1 is the lower quartile
Area to the left of Q_1D = Area Q_1Q_2BD = Area Q_2Q_3EB
= Area to right of Q_3E
$\frac{1}{4}$ Area under the distribution curve.

Fig. 5.1 Measures of location and dispersion

b. **Standard Deviation** (s) Standard deviation is defined as 'The square root of the mean of the squares of the deviations of the observations from their mean', i.e., standard deviation is an r.m.s. value; it is the root mean square of the differences between the observations and the mean.

Now: Deviation from the mean $= x - \bar{x}$.

Therefore, mean square deviation from the mean

$$= \frac{1}{N}\Sigma f(x-\bar{x})^2$$

and standard deviation

$$s = \sqrt{\left[\frac{1}{N}\Sigma f(x-\bar{x})^2\right]} \tag{5.6}$$

(It is often more convenient to consider s^2 and this is called *variance*.)

$$\text{Variance} = (\text{Standard Deviation})^2.$$

c. **Mean Deviation About the Mean** Care must be taken not to confuse the much used 'standard deviation' (defined above) with the little used 'mean deviation about the mean'—often loosely called just 'mean deviation'. This is the mean of the numerical values of the deviations of the observations from the mean (ignoring signs).

The symbol for numerical difference is $|x-\bar{x}|$ (e.g., $3-5 = -2$ but $|3-5| = 2$) and is called the *modulus* of $x-\bar{x}$. Hence:

$$\text{Mean deviation from the mean} = \frac{1}{N}\Sigma f(|x-\bar{x}|) \qquad (5.7)$$

d. **Coefficient of Variation** (v) This is a non-dimensional measure of dispersion and is defined as

$$v = \frac{s}{\bar{x}} \qquad (5.8)$$

$$\left(\text{or, as a percentage, } \frac{100s}{\bar{x}}\%\right)$$

5.5 Calculation of mean and standard deviation

In order to reduce the magnitude of the numbers involved in calculating the mean ($\bar{x}$) and standard deviation (s), and thereby to make the arithmetic easier (hence reducing the likelihood of errors), it is often advisable first to 'code' the data, that is, to subtract a suitable constant from all values of the variate then divide by another suitable constant. Coding is used to simplify the numerical values of the variate. Consider, for instance, the 'class mid-marks' x of the example which follows on page 57.

3·752 3·757 3·762 3·767 3·772 3·777 3·782

First subtract 3·767 from each value, producing

−0·015 −0·010 −0·005 0 +0·005 +0·010 +0·015

Now divide each value by 0·005, producing the 'coded data' u.

−3 −2 −1 0 +1 +2 +3

It will be readily obvious that the values of u are much easier to work with than the original values of x.

If the original values of the variate are anything other than the simplest whole numbers then there is everything to gain, and nothing to lose, from 'coding' the data, and the following set of formulae should be used.

$$\text{Coded data: } u = \frac{x-A}{c} \qquad (5.9)$$

where A and c are arbitary constants; A being taken—for preference —equal to a value of the variate near to the mode, and c being equal to the class interval (i.e., the difference between any two successive class mid-marks). Then

$$\bar{u} = \frac{1}{N}\Sigma fu \text{ where } N = \Sigma f \tag{5.10}$$

and

$$\bar{x} = A + c.\bar{u} \tag{5.11}$$

$$\text{also sample variance } s^2 = c^2\left[\frac{\Sigma fu^2}{N} - (\bar{u})^2\right] \tag{5.12}$$

a. **Example**

Shaft diameters (in.) Class boundaries	Class Mid-mark in. x	Number of Shafts (Frequency) f	Coded data u	fu	fu^2
3·7495–3·7545	3·752	2	−3	− 6	18
3·7545–3·7595	3·757	5	−2	−10	20
3·7595–3·7645	3·762	16	−1	−16	16
3·7645–3·7695	3·767	59	0	0	0
3·7695–3·7745	3·772	11	1	11	11
3·7745–3·7795	3·777	6	2	12	24
3·7795–3·7845	3·782	1	3	3	9
Totals		100		+26 −32 = −6	98

Coded data: $u = \dfrac{x-A}{c}$
where $A = 3{\cdot}767$
and $c = 0{\cdot}005$

$$N = \Sigma f = 100$$
$$\Sigma fu = -6$$
$$\Sigma fu^2 = 98$$
$$\bar{u} = \frac{1}{N}\Sigma fu = -0{\cdot}06$$
$$\bar{x} = A + c\bar{u} = 3{\cdot}767 + (0{\cdot}005)(-0{\cdot}06)$$
$$= 3{\cdot}767 - 0{\cdot}0003 = 3{\cdot}7667 \text{ in.}$$
$$s^2 = c^2\left[\frac{1}{N}\Sigma fu^2 - (\bar{u})^2\right] = (0{\cdot}005)^2[0{\cdot}98 - (-0{\cdot}06)^2]$$
$$= (0{\cdot}005)^2[0{\cdot}98 - 0{\cdot}0036]$$

i.e., $s^2 = (0{\cdot}005)^2(0{\cdot}9764)$

therefore $s = 0{\cdot}005 \times 0{\cdot}9881$

$= 0{\cdot}004\,940\,5$ in.

Mean diameter 3·7667 in.
Standard deviation of diameter 0·00494 in.

It will be readily appreciated that this method of using 'coded data' is essential if the calculations are to be carried out without the help of mechanical aids (calculating machines or computers). If such aids are available then it may be possible to use the direct formulae based on the definitions:

$$\bar{x} = \frac{1}{N}\Sigma fx \tag{5.4}$$

$$s = \sqrt{\left[\frac{1}{N}\Sigma f(x-\bar{x})^2\right]} \tag{5.6}$$

although even in such cases the derived form

$$s = \sqrt{\left[\frac{1}{N}(\Sigma fx^2)-(\bar{x})^2\right]} \tag{5.13}$$

will probably be simpler to use, and less liable to errors due to 'rounding off'.

Proofs of the derived expressions are given in the notes at the end of this chapter. It should be remembered that they are exact, and are not approximate forms. It should also be noted that the value of $\bar{x}$ for a given set of data, is independent of the values chosen for the constants A and c in defining u.

b. **Worked example** Determine the mean and standard deviation of the contents of the 40 bottles listed in section 4.5.

Class Mid-mark x	Class Frequency f	$\frac{x-4744{\cdot}5}{10} = u$	fu	fu^2
4704·5	1	−4	−4	16
4714·5	3	−3	−9	27
4724·5	4	−2	−8	16
4734·5	8	−1	−8	8
4744·5	10	0	0	0
4754·5	7	1	7	7
4764·5	3	2	6	12
4774·5	2	3	6	18
4784·5	2	4	8	32
	40		27 −29 = −2	136

$$\Sigma f = 40$$
$$\Sigma fu = -2$$
$$\Sigma fu^2 = 136$$

$$\bar{u} = \frac{\Sigma fu}{\Sigma f} = -\frac{2}{40} = -0{\cdot}05$$

Now $$u = \frac{x-4744{\cdot}5}{10}$$

therefore $$\bar{x} = 10\bar{u}+4744{\cdot}5$$
$$= 4744$$

i.e., Mean contents = 4744 cm^3.

$$s_u = \sqrt{\left(\frac{\Sigma fu^2}{\Sigma f}-\bar{u}^2\right)}$$
$$= \sqrt{\left(\frac{136}{40}-0{\cdot}0025\right)}$$
$$= \sqrt{(3{\cdot}4-0{\cdot}0025)}$$
$$= \sqrt{3{\cdot}3975} = 1{\cdot}8433$$

Now $$s_x = 10s_u = 18{\cdot}43$$

i.e., Standard deviation of contents = 18·4 cm^3.

Similar calculations based on a sample taken from another filling machine, working to the same specification, showed a mean contents of 4754 cm^3 and a standard deviation of 14·5 cm^3. This shows that the second machine has not only been set up to fill more nearly to the specified 4·750 litres, but also that it is less variable from bottle to bottle (shown by the smaller value of standard deviation).

5.6 Proof of formulae used in calculating $\bar{x}$ and s

By definition

$$\bar{x} = \frac{1}{N}\Sigma fx \tag{5.4}$$

Let $u = \dfrac{x-A}{c}$ where A and c are constants, then $x = A+cu$.

Hence $$\bar{x} = \frac{1}{N}\Sigma f(A+cu)$$
$$= \frac{1}{N}[\Sigma fA+\Sigma fcu]$$
$$= \frac{1}{N}[A\Sigma f+c\Sigma fu]$$
$$= \frac{1}{N}[NA+c\Sigma fu] \text{ since } N = \Sigma f$$
$$= A+c.\frac{1}{N}\Sigma fu$$
$$= A+c\bar{u} \text{ where } \bar{u} = \frac{1}{N}\Sigma fu.$$

i.e., $$\bar{x} = A+c\bar{u} \tag{5.11}$$

Again, by definition,

$$s = \sqrt{\left[\frac{1}{N}\Sigma f(x-\bar{x})^2\right]} \tag{5.6}$$

i.e.,

$$\begin{aligned} Ns^2 &= \Sigma f[x^2 - 2x\bar{x} + (\bar{x})^2] \\ &= \Sigma fx^2 - 2\bar{x}\Sigma fx + (\bar{x})^2\Sigma f \\ &= \Sigma fx^2 - 2\bar{x}(N\bar{x}) + N(\bar{x})^2 \\ &= \Sigma fx^2 - 2N(\bar{x})^2 + N(\bar{x})^2 \\ &= \Sigma fx^2 - N(\bar{x})^2 \end{aligned}$$

therefore

$$s^2 = \frac{1}{N}\Sigma fx^2 - (\bar{x})^2$$

i.e.,

$$s = \sqrt{\left[\frac{1}{N}(\Sigma fx^2) - (\bar{x})^2\right]} \tag{5.13}$$

Now $x = A + cu$

and $\bar{x} = A + c\bar{u}$

therefore $x - \bar{x} = c(u - \bar{u})$

Hence

$$\begin{aligned} Ns^2 &= \Sigma fc^2(u-\bar{u})^2 \\ &= c^2\Sigma f(u-\bar{u})^2 \\ &= c^2\Sigma f[u^2 - 2u\bar{u} + (\bar{u})^2] \\ &= c^2[\Sigma fu^2 - 2\bar{u}\Sigma fu + (\bar{u})^2\Sigma f] \\ &= c^2[\Sigma fu^2 - N(\bar{u})^2] \end{aligned}$$

i.e.,

$$s^2 = c^2\left[\frac{1}{N}(\Sigma fu^2) - (\bar{u})^2\right] \tag{5.12}$$

5.7 Problems

1. A hardness check on 500 pieces of nickel-steel yielded the following results:

Brinell Hardness Number (Inclusive values, to the nearest whole number)	Number of pieces
290–294	17
295–299	86
300–304	240
305–309	104
310–314	43
315–319	10

Compute the mean and standard deviation of Brinell Hardness Number for these 500 test pieces.

2. The production of a radar component is checked by examining samples of 4. The table shows the number of defectives found in 200 samples.

No. of defectives	0	1	2	3	4
No. of samples	62	85	40	11	2

Calculate
(a) the mean number of defectives
(b) the standard deviation of the number of defectives.
(c) the median.
(d) the mode.

3. The number of stoppages on 400 consecutive shifts in an industrial firm are recorded below. Calculate the mean and standard deviation of the number of stoppages per shift.

Stoppages per shift	0	1	2	3	4	5
No. of shifts	245	119	30	4	2	0

4. The table shows the distribution of the diameters of the heads of rivets manufactured by a certain company. Compute
(a) the mean diameter
(b) standard deviation of diameter
(c) the median

Inclusive Diameters (in.)	Number of Rivets
0·7247–0·7249	2
0·7250–0·7252	6
0·7253–0·7255	8
0·7256–0·7258	15
0·7259–0·7261	42
0·7262–0·7264	68
0·7265–0·7267	49
0·7268–0·7270	25
0·7271–0·7273	18
0·7274–0·7276	12
0·7277–0·7279	4
0·7280–0·7282	1

5. Use the data of problem 4 of chapter 4 to compute the mean and standard deviation of diameter for the given sample of 60 ball bearings.

6. The table shows the distribution of lifetime of radio valves tested at the A and B Co. laboratory.

Lifetime (hours)	No. of tubes
300– 400$^-$	14
400– 500$^-$	46
500– 600$^-$	58
600– 700$^-$	76
700– 800$^-$	68
800– 900$^-$	62
900–1000$^-$	48
1000–1100$^-$	22
1100–1200$^-$	6

Calculate
(a) the mean lifetime
(b) the standard deviation of lifetime
(c) coefficient of variation
(d) interquartile range.

7. Tests made on electrical contacts to determine the number, out of sets of 12, which failed to open (due to welding) before a reversal rate of 1000 cycles per minute had been reached, yielded the following results:

No. of contacts welded	0	1	2	3	4	5
Frequency	10	13	13	8	4	2

Calculate
(a) the mean,
(b) the variance,
(c) the mean deviation from the mean of the number of contacts failing to open.
Determine also
(d) the median, and
(e) the mode.

6. Special distributions

Depending upon the circumstances giving rise to the formation of the data, their distribution may conform to one or other of certain standard patterns, whose properties are known. We will consider the conditions which give rise to, and the properties of, four such distributions: the Hypergeometric, Binomial, Poisson, and Normal. The relationships between these distributions, and their chief properties, are summarized in table form at the end of the chapter.

6.1 Hypergeometric distribution

Consider a random sample of n pieces taken from a finite batch containing a total of N pieces, of which a proportion p are defective.

Number of defective pieces in batch	$= Np$
Number of good pieces in batch	$= N(1-p)$
	$= Nq$ where $q = 1-p$
Number of ways of choosing x defectives from Np	$= \binom{Np}{x}$
Number of ways of choosing $(n-x)$ good pieces from Nq	$= \binom{Nq}{n-x}$
Number of ways of choosing any n pieces from N	$= \binom{N}{n}$

Therefore, probability of obtaining a sample of n pieces containing x defectives $= P(x)$ where

$$P(x) = \frac{\text{No. of ways of obtaining } x \text{ defectives and } (n-x) \text{ good pieces}}{\text{No. of ways of obtaining any } n \text{ pieces from the batch of } N \text{ pieces}}$$

i.e.,

$$P(x) = \frac{\binom{Np}{x}\binom{Nq}{n-x}}{\binom{N}{n}} \qquad (6.1)$$

For a given batch N, p, and q are fixed; for a given sampling scheme n is fixed and hence probability $P(x)$ depends only on the number of defectives (x) in the sample.

Hence eq. (6.1) describes the probability distribution and is known as 'The Hypergeometric probability distribution function'.

a. **Mean and Variance** If many random samples, each of size *n*, are taken from the same population, the numbers of defectives found will differ from sample to sample. These 'numbers of defectives per sample' will themselves form a distribution whose mean and variance can be determined. It can be shown that:

Mean number of defectives ($\bar{x}$) in sample of n:

$$\bar{x} = np \tag{6.2}$$

Variance of number of defectives (σ^2) in sample of n:

$$\sigma^2 = npq\left(\frac{N-n}{N-1}\right) \tag{6.3}$$

b. **Example** Determine the probability that a batch of 150 pieces containing 8 defectives will yield a sample of 10 pieces of which exactly 3 are defective.

No. of defective pieces	$= Np$	=	8
No. of good pieces	$= Nq$	=	142
No. of pieces in batch	$= N$	=	150
No. of pieces in sample	$= n$	=	10
No. of defectives in sample	$= x$	=	3.

Probability of x defectives in sample of n pieces

$$= P(x) = \frac{\binom{Np}{x}\binom{Nq}{n-x}}{\binom{N}{n}}$$

Now $$\binom{N}{n} = \frac{N(N-1)(N-2)\ldots\text{to } n \text{ terms}}{n(n-1)(n-2)\ldots3\times2\times1}$$

$$\binom{150}{10} = \frac{150\times149\ldots142\times141}{10\times9\times8\ldots3\times2\times1}$$

$$\binom{Np}{x} = \binom{8}{3} = \frac{8\times7\times6}{3\times2\times1}$$

$$\binom{Nq}{n-x} = \binom{142}{7} = \frac{142\times141\times140\times\ldots\times136}{7\times6\times5\times\ldots\times2\times1}$$

$$P(3) = \frac{\binom{8}{3}\times\binom{142}{7}}{\binom{150}{10}} = 0{\cdot}0095$$

That is, the probability of obtaining a random sample of 10 pieces containing exactly 3 defectives (from a batch of 150 pieces of which 8 are defective) is 0·0095.

6.2 Binomial distribution

When the batch size is big enough, in relation to the proportion defective and the sample size, to permit the assumption that the probability of drawing a defective piece is the same at the end of the sample as it was at the beginning, then we can state:

$$P(x) = \binom{n}{x} p^x q^{n-x} \tag{6.4}$$

This is the $(x+1)$th term in the expansion of the binomial $(q+p)^n$ in ascending powers of p and such probabilities are said to form the Binomial probability distribution.

In particular:

$$P(0) = q^n$$
$$P(1) = \binom{n}{1} q^{n-1} p = nq^{n-1}p$$
$$P(2) = \binom{n}{2} q^{n-2} p^2 = \frac{n(n-1)}{2} q^{n-2} p^2$$
$$P(3) = \binom{n}{3} q^{n-3} p^3 = \frac{n(n-1)(n-2)}{3\times 2} q^{n-3} p^3 \quad \text{etc.}$$

But, by the Binomial Theorem

$$(q+p)^n \equiv q^n + nq^{n-1}p + \frac{n(n-1)}{2} q^{n-2}p^2 + \frac{n(n-1)(n-2)}{3\times 2} q^{n-3}p^3 + \text{etc.}$$

$$P(0)+P(1) \quad +P(2) \qquad +P(3) \qquad + \text{etc.}$$

i.e., the probabilities of obtaining a random sample of n pieces, from a large batch of which a proportion p are defective, containing 0, 1, 2, 3, etc., defectives, are the successive terms of the expansion of $(q+p)^n$ in ascending powers of p.

a. **Mean and Variance** For the Binomial distribution, the probability of obtaining exactly x successes in n trials is

$$P(x) = \binom{n}{x} q^{n-x} p^x$$

It can be shown that:
The mean number of successes in n trials

$$\bar{x} = np \tag{6.5}$$

variance of number of successes

$$\sigma^2 = npq \tag{6.6}$$

b. **Example** A quality control scheme requires that random samples of 5 pieces be taken from the production at regular intervals.

How often would you expect the sample to contain more than 1 defective if the process defective rate is

(a) 10 per cent
(b) 30 per cent.

p = probability that any piece taken from production will be defective
= process proportion defective
n = number of pieces in sample.
$P(0)$ = Probability of obtaining sample containing 0 defectives.
$P(1)$ = Probability of obtaining sample containing 1 defective.
P = Probability of obtaining sample containing more than 1 defective.
$P = 1-[P(0)+P(1)]$.
$P(0) = q^n$ where $q = 1-p$
$P(1) = nq^{n-1}p$.

(a) $n = 5$ $p = 0{\cdot}1$. From Table A

	$P(0) = (0{\cdot}9)^5$	$= 0{\cdot}5905$
	$P(1) = 5(0{\cdot}9)^4(0{\cdot}1)$	$= 0{\cdot}9185-0{\cdot}5905 = 0{\cdot}3280$
therefore	$P(0)+P(1)$	$= 0{\cdot}9185$
and	P	$= 0{\cdot}0815$

i.e., we would expect about 1 in 12 samples to contain more than 1 defective.

(b) $n = 5$ $p = 0{\cdot}3$. From Table A

	$P(0) = (0{\cdot}7)^5$	$= 0{\cdot}1681$
	$P(1) = 5(0{\cdot}7)^4(0{\cdot}3)$	$= 0{\cdot}5282-0{\cdot}1681 = 0{\cdot}3601$
therefore	$P(0)+P(1)$	$= 0{\cdot}5282$
and	P	$= 0{\cdot}4718$

i.e., we would expect about 5 out of 10 samples to contain more than 1 defective.

6.3 Poisson distribution

When the number of possible 'defects' per unit is unlimited and the probability (p) of any one defect happening is very small, then the probability of x 'defects' in a unit is $P(x)$ and is given by

$$P(x) = \frac{e^{-\lambda}\lambda^x}{x!} \tag{6.7}$$

'e' is the exponential function (e = 2·7183 approx.)
λ is the average number of 'defects' per unit.

In practice, provided n is large and p is small, Poisson probabilities give a good approximation to Binomial probabilities, by using

$$\lambda = np.$$

Remember, however, that although the Poisson distribution is useful as an approximation to the Binomial it is also a distribution in its own right. It may be said that the Poisson distribution is the distribution of unlikely events.

As an approximation to the Binomial it has the advantage of easy calculation and each term can be generated readily from the preceding term.

$$P(x) = \frac{e^{-\lambda}\lambda^x}{x!}$$

$$P(0) = \frac{e^{-\lambda}\lambda^0}{0!} = e^{-\lambda}$$

(N.B. $a^0 = 1$ and $0! = 1$; the latter by definition.)

$$P(1) = \frac{e^{-\lambda}\lambda^1}{1!} = \lambda e^{-\lambda} = \lambda P(0).$$

$$P(2) = \frac{e^{-\lambda}\lambda^2}{2!} = \frac{\lambda}{2}\,\frac{e^{-\lambda}\lambda}{1!} = \frac{\lambda}{2}P(1)$$

$$P(3) = \frac{e^{-\lambda}\lambda^3}{3!} = \frac{\lambda}{3}\,\frac{e^{-\lambda}\lambda^2}{2!} = \frac{\lambda}{3}P(2)$$

etc.

and in general:

$$P(x) = \frac{\lambda}{x}P(x-1) \qquad (6.8)$$

a. **Mean and Variance** It is a property of the Poisson distribution that the mean and variance of number of events per interval are numerically equal.

$$\text{Mean } (\bar{x}) = \text{Variance } (\sigma^2) = \lambda \qquad (6.9)$$

b. **Example 1** A machine shop employing a large number of men finds that, over a period of time, the average absentee rate is 3 men per shift. Calculate the probability that, on a given shift,

(a) exactly 2 men will be absent,

(b) more than 4 men will be absent.

$$P(x) = \frac{e^{-\lambda}\lambda^x}{x!}$$

λ = average number of absentees per shift.
= 3.

$e^{-\lambda} = e^{-3} = 0{\cdot}0498$ (from tables).

(a) $$P(2) = \frac{e^{-\lambda}\lambda^2}{2!}$$

$$= \frac{0{\cdot}0498 \times 3^2}{2 \times 1} = \frac{0{\cdot}0498}{2} \times 9 = 0{\cdot}2241.$$

i.e., probability that exactly 2 men will be absent $= 0{\cdot}2241$.

(b) $P(0) = e^{-\lambda} = 0{\cdot}0498$

$P(1) = \lambda P(0) = 0{\cdot}1494$

$P(2) = \frac{\lambda}{2} P(1) = 0{\cdot}2241$

$P(3) = \frac{\lambda}{3} P(2) = 0{\cdot}2241$

$P(4) = \frac{\lambda}{4} P(3) = 0{\cdot}1681$

$P(0)+P(1)+P(2)+P(3)+P(4) = 0{\cdot}8555$

Therefore, probability that more than 4 men will be absent

$= 1 - 0{\cdot}8555$
$= 0{\cdot}1445$

c. **Example 2. (As an approximation to the Binomial)** A random sample of 50 pieces is taken from a batch containing 2 per cent defectives.

Calculate the probability that the sample will contain less than 3 defectives.

n = sample size = 50
p = proportion defective = 0·02
λ = average number of defectives per sample = np
$= 50 \times 0{\cdot}02 = 1.$

$$P(x) = \frac{e^{-\lambda}\lambda^x}{x!}$$

$e^{-\lambda} = e^{-1} = 0{\cdot}3679$

$P(0) = e^{-\lambda} = 0{\cdot}3679$

$P(1) = \lambda P(0) = 0{\cdot}3679$

$P(2) = \frac{\lambda}{2} P(1) = 0{\cdot}18395$

$P(0)+P(1)+P(2) = 0{\cdot}91975$

so the probability that the sample will contain less than 3 defectives is 0·9198.

(This means that we can expect 92 out of 100 samples from such batches to contain less than 3 defectives.)

N.B. These figures could have been obtained without calculation by the use of the cumulative Poisson probabilities given in Table B.

6.4 Normal (or Gaussian) distribution

The three distributions so far considered all relate to discrete data, this fourth type describes continuous data. Many naturally occurring distributions, e.g., heights of adult males, weights of eggs, diameters of nominal $\frac{1}{2}$-in. diameter ball bearings, etc., conform to approximately the same shape. This characteristic distribution is bell-shaped and symmetrical.

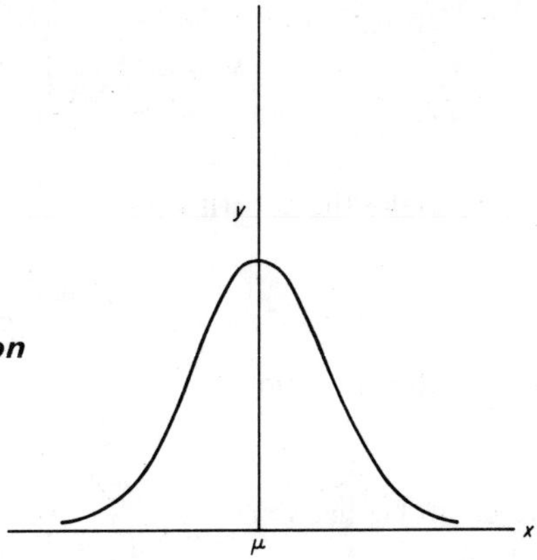

Fig. 6.1 Normal distribution

The equation of this curve is

$$y = \frac{1}{\sigma\sqrt{(2\pi)}} e^{-(x-\mu)^2/2\sigma^2} \qquad (6.10)$$

where x can take any value from minus infinity to plus infinity, and it can be shown that:

μ is the mean value of x

and σ^2 is the variance of x.

Just as the area of a histogram column represents probability, so also the probability that a random value of x (from a given Normal distribution whose mean is μ and variance σ^2) will have a value between x_1 and x_2 is represented by the area under the Normal curve between these limits.

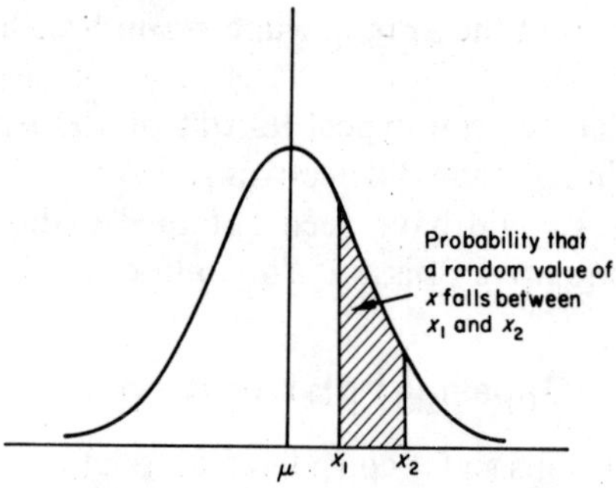

Fig. 6.2 Area under Normal curve

The probability that a random value of x lies between x_1 and x_2 is P, where

$$P = \frac{1}{\sigma\sqrt{(2\pi)}}\int_{x_1}^{x_2} e^{-(x-\mu)^2/2\sigma^2}\,dx \qquad (6.11)$$

If we make the substitution

$$Z = \frac{x-\mu}{\sigma} \qquad (6.12)$$

then eq. (6.10) becomes

$$y = e^{-\frac{1}{2}z^2}$$

and eq. (6.11) becomes

$$P = \int_{z_1}^{z_2} e^{-\frac{1}{2}z^2}\,dz$$

This substitution reduces every Normal distribution to a common standard and the equations produced are said to be those of the 'standardized Normal curve', whose mean is zero and variance unity.

In other words, the 'standardized' Normal variate (z) is distributed about zero mean with unit variance.

Thus x has mean μ and variance σ^2
but z has mean 0 and variance 1.

Areas under the standardized Normal curve have been evaluated and are set out in Table C.

In most practical problems the values of the population mean μ and variance σ^2 are unknown. In their absence we must use sample mean $\bar{x}$ and variance s^2 as the best available estimates. Provided the samples are not small these approximations will be acceptably accurate. The substitution

$$z = \frac{x-\mu}{\sigma}$$

then becomes

$$z = \frac{x-\bar{x}}{s}$$

a. **Table of Areas Under the Normal Curve** Since the areas represent probabilities the total area under the curve is taken as 1 unit.

The curve is symmetrical so the area to each side of the mean is 0·5, i.e., the probability of obtaining a value of x greater than the mean (μ) is 0·5.

Table C gives the probability that $z(z = (x-\mu)/\sigma)$ will lie between 0 and $z_1(z_1 = (x_1-\mu)/\sigma)$. Hence, to find the probability that z will lie between z_1 and z_2 it is necessary to read two values from the table and the required probability is given by the difference between these values.

Again, the probability that z will exceed z_1 is found by subtracting the table value corresponding to z_1, from 0·5.

Since the curve is symmetrical the probability of obtaining a value of z between 0 and $-z_1$ (i.e., x_1 is less than μ) is equal to the probability of obtaining a value of z between 0 and $+z_1$. Accordingly, the table of areas has only been printed for the upper half (or 'tail') of the curve.

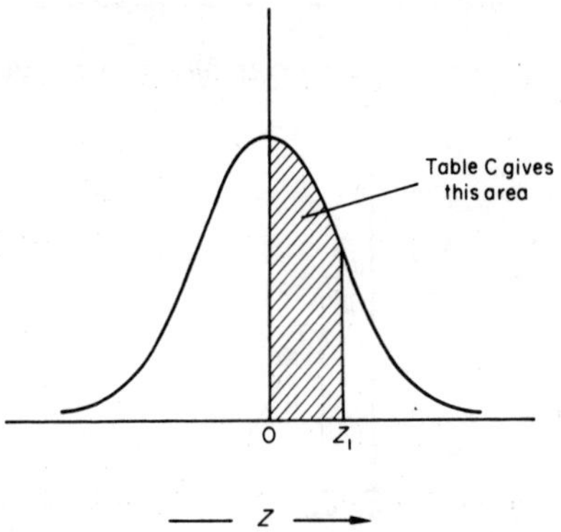

Fig. 6.3 Area under Normal curve

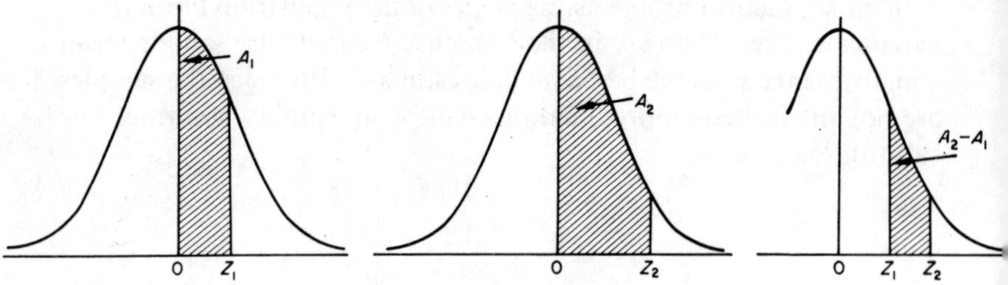

Fig. 6.4 Area under Normal curve

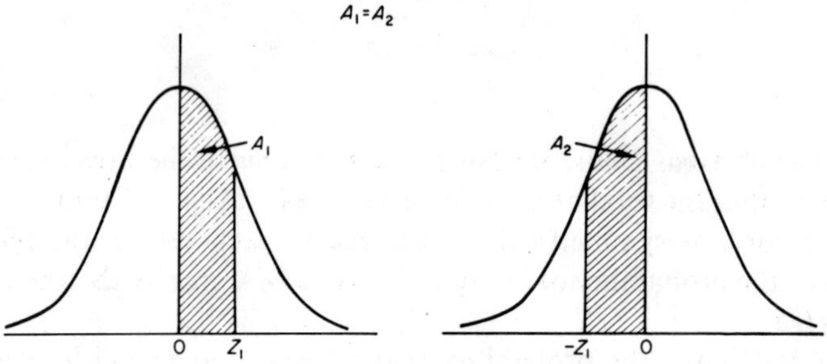

Fig. 6.5 Area under Normal curve

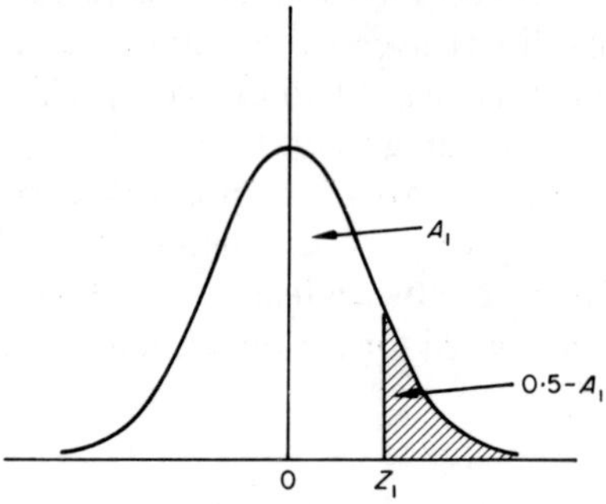

Fig. 6.6 Area under Normal curve

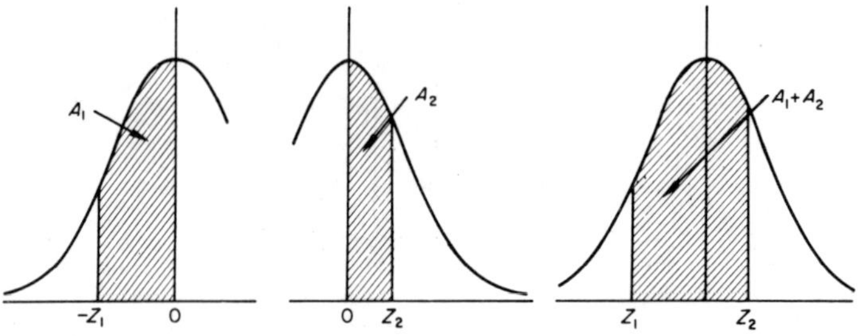

Fig. 6.7 Area under Normal curve

b. **Process Capability** From the Table C we see that, when

$$z_1 = 3{\cdot}0 \quad A_1 = 0{\cdot}49865$$

$$\text{probability of } z \text{ exceeding } 3{\cdot}0 = 0{\cdot}5 - 0{\cdot}49865$$
$$= 0{\cdot}00135$$

and probability of z less than $-3{\cdot}0 = 0{\cdot}00135$.
therefore probability of z being outside the band $\pm 3 = 0{\cdot}0027$

But
$$z = \frac{x-\mu}{\sigma}$$

therefore
$$x = \mu + \sigma z$$

Hence probability of x being outside the band $\mu \pm 3\sigma$

$$= 0{\cdot}0027,$$

i.e., from a Normal Population, only 0·27 per cent differ from the mean by more than 3 standard deviations.

This means that, if the variations (from piece to piece) of a given dimension on a manufactured component are Normally distributed about mean μ with standard deviation σ, then only about 3 in every 1000 pieces produced will have the given dimension outside the limits $\mu \pm 3\sigma$. Consequently, these are the limits of accuracy to which the manufacturing process is capable of consistently working, that is:

$$\text{Process Capability} = \text{Mean} \pm 3 \times \text{standard deviation} \qquad (6.13)$$

If the drawing limits are $\pm T$ say, then if T is less than 3σ the production of pieces 'not to drawing' is unavoidable, and the percentage of such pieces to be expected can be calculated. If T is greater than 3σ then only a very occasional piece will not conform to specification.

It must be emphasized that Process Capability (eq. 6.13) refers only to *random* variation from piece to piece. Any *systematic* change due to assignable causes, e.g., a shift in the position of a 'stop' or wear at a tool point, must be allowed for separately.

c. **Example 1** A machine is producing shafts to a nominal diameter of 2·000 in. with tolerance $\pm 0{\cdot}005$ in. It is found that the standard deviation of the diameters produced by this machine is 0·001 in. Measurements show that wear at the tool point causes an increase in the diameter produced of 0·001 in. for every 18 shafts machined.

Calculate
(a) the initial diameter the operator should aim to set his machine to produce, and
(b) the number of shafts you would expect to be able to produce before re-setting the tool stop,

if the latter is to be a maximum consistent with negligible chance of producing shafts outside limits.

(a) Aiming point must be 3σ above lower drawing limit, that is

$$(2{\cdot}000-0{\cdot}005)+0{\cdot}003 = 2{\cdot}000-0{\cdot}002 \text{ in.}$$

(b) He can allow a 'drift' due to tool wear up to 3σ below upper drawing limit, that is

$$(2{\cdot}000+0{\cdot}005)-0{\cdot}003 = 2{\cdot}000+0{\cdot}002 \text{ in.}$$

Therefore total drift = 0·004 in.

Now drift = 0·001 in. over 18 shafts.

Therefore number of shafts machined between re-sets = $4 \times 18 = 72$.

d. **Example 2** A bore in a mass-produced part is found to have a mean diameter of 0·2498 in. with a standard deviation of 0·0012 in. Determine the percentage of the pieces produced you would expect to lie outside the drawing limits of $0{\cdot}250 \pm 0{\cdot}002$ in.

$$\mu = 0{\cdot}2498 \text{ in.}$$
$$\sigma = 0{\cdot}0012 \text{ in.}$$

$$x_1 = 0{\cdot}250-0{\cdot}002 \qquad z_1 = \frac{x_1-\mu}{\sigma} = \frac{0{\cdot}2480-0{\cdot}2498}{0{\cdot}0012} = -1{\cdot}5$$

$$x_2 = 0{\cdot}250+0{\cdot}002 \qquad z_2 = \frac{x_2-\mu}{\sigma} = \frac{0{\cdot}2520-0{\cdot}2498}{0{\cdot}0012} = +1{\cdot}83.$$

From Table C:

$$A_1 = 0{\cdot}4332$$
$$A_2 = 0{\cdot}4664$$
$$A_1+A_2 = 0{\cdot}8996$$

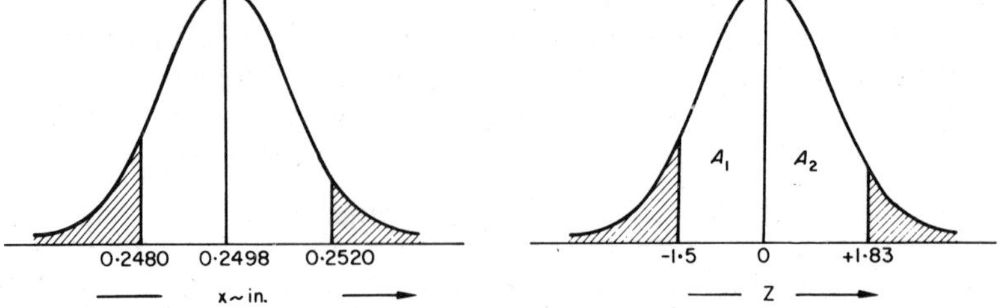

Fig. 6.8

Thus percentage within drawing limits = 89·96 so 10·04 per cent would be expected to not conform to drawing.

6.5 Use of Normal distribution as an approximation to the Binomial

For the Binomial distribution, $(q+p)^n$, the histogram will be made up of so many columns when n is very large that it will be approximately a smooth curve. Since the calculation of Binomial probabilities is an arduous task it is frequently useful to obtain an approximation by employing the comparable Normal distribution. By 'comparable' we imply that it must have the same mean and variance as the Binomial distribution in question.

a. **Example** Determine the probability that a random sample of 80 pieces, taken from a batch containing 5 per cent defectives, will contain less than 3 defectives.

$$n = 80$$
$$p = 0{\cdot}05$$
$$q = 1-p = 0{\cdot}95$$
$$\mu = \text{mean} = np = 80 \times 0{\cdot}05 = 4{\cdot}0$$
$$\sigma^2 = \text{variance} = npq = 4{\cdot}0 \times 0{\cdot}95 = 3{\cdot}80$$

Therefore $\sigma = 1{\cdot}949$

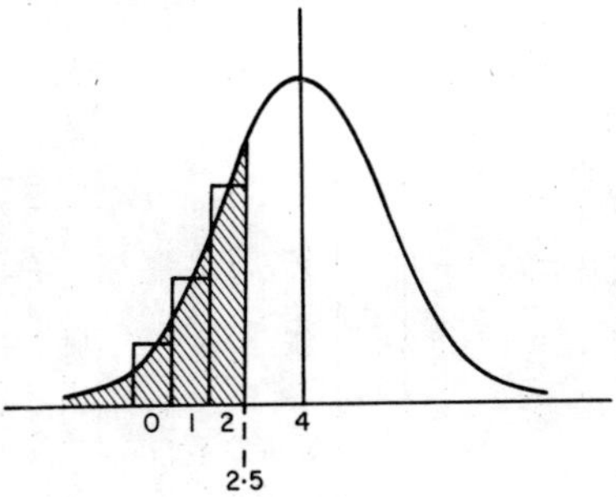

Fig. 6.9

For Normal probability approximating to Binomial probability of less than 3 we require the area to the left of $x = 2{\cdot}5$. We use $x_1 = 2{\cdot}5$, not $x = 3$, since we require an approximation to the areas of the binomial histogram columns for 0, 1, and 2 defectives, and the '2' column is drawn 1 unit wide from 1·5 to 2·5.

This shift of $\frac{1}{2}$ unit is known as the continuity correction and is most important. It must always be used when obtaining an approximation to a discrete distribution by means of a continuous distribution.

Table 6.1 **Relationship between the distributions**

		Distributions used as approximations to Hypergeometric		
Distribution	1 Hypergeometric (Discrete Data)	2 Binomial (Discrete Data)	3 Poisson (Discrete Data)	4 Normal (Continuous Data)
Probability Function	$P(x) = \frac{\binom{Np}{x}\binom{Nq}{n-x}}{\binom{N}{n}}$ where $q = 1 - p$	$P(x) = \binom{n}{x}p^x q^{n-x}$	$P(x) = \frac{e^{-\lambda}\lambda^x}{x!}$ where $\lambda = np$	$dP = \frac{1}{\sigma\sqrt{(2\pi)}} e^{-(x-\mu)^2/2\sigma^2} dx$
Mean	np	np	λ	μ
Variance	$npq\left(\frac{N-n}{N-1}\right)$	npq	λ	σ^2
Conditions for validity of approximation	Theoretically correct in all cases	N, n, p such that p is not measurably altered by sampling	As for 2, but also $p \leqslant 0{\cdot}1$ $np < 5$	As for 2, but also $p \geqslant 0{\cdot}1$ $np > 5$
Comments	Computation is very involved and this distribution should only be used in exceptional circumstances.	Computation is again difficult and either 3 or 4 should be used whenever possible.	Used when n is large and p is small.	Used when n is large and p *not* very small.

N = Number of pieces in batch. n = Number of pieces in sample. p = Proportion of defective pieces in batch.
$P(x)$ = Probability that a random sample will contain exactly x 'defectives'.

The direction of the correction will depend on the question. If we had required 'the probability of 3 or less ...' then we would have had to use $x_1 = 3{\cdot}5$. Each case must be treated on its merits.

In the present case:

$$x_1 = 2{\cdot}5$$

$$z_1 = \frac{x_1 - \mu}{\sigma} = \frac{2{\cdot}5 - 4{\cdot}0}{1{\cdot}949}$$

$$= -0{\cdot}7696.$$

$$A_1 = 0{\cdot}2793 \text{ from tables.}$$

$$\text{Probability of } x \text{ less than } 2{\cdot}5 = 0{\cdot}5 - 0{\cdot}2793 = 0{\cdot}2207$$

That is, the probability of obtaining a random sample of 80 in which less than 3 are defective is 22 per cent.

6.6 Worked example

In the worked examples of the two previous chapters (pages 47 and 58) we have investigated the distribution of contents of bottles filled by two machines, using an automatic process. We found

	cm³	
Filling Machine	Mean	Standard Deviation
A	4744	18·4
B	4754	14·5

The bottles contain a nominal 4·75 litres but are warranted to contain not less than 4·70 litres. Use the available data to decide if the two machines are functioning satisfactorily, and if not, state what corrective action should be taken. Assume both distributions to be Normal.

Process capability = mean $\pm 3 \times$ standard deviation. Hence

Machine	Mean	3 × Standard Deviation	Process capability limits cm³
A	4744	55·2	4688·8 to 4799·2
B	4754	43·5	4710·5 to 4797·5

This shows that only about 1 bottle in 1000 filled by machine B will contain less than 4710·5 cm³ so the warranted minimum of 4·70 litres is most unlikely to be transgressed. In fact an economy could be brought about by reducing the average 'fill' by some 4 cm^3, thus reducing the mean from 4754 cm^3 to the nominal 4750 cm^3.

Machine A is another matter altogether! The process capability lower limit of 4688·8 cm^3 is well below the warranted minimum. To determine the percentage of bottles likely to contain less than 4700 cm^3 we must find the area under the Normal curve to the left of this value. (See Fig. 6.10.)

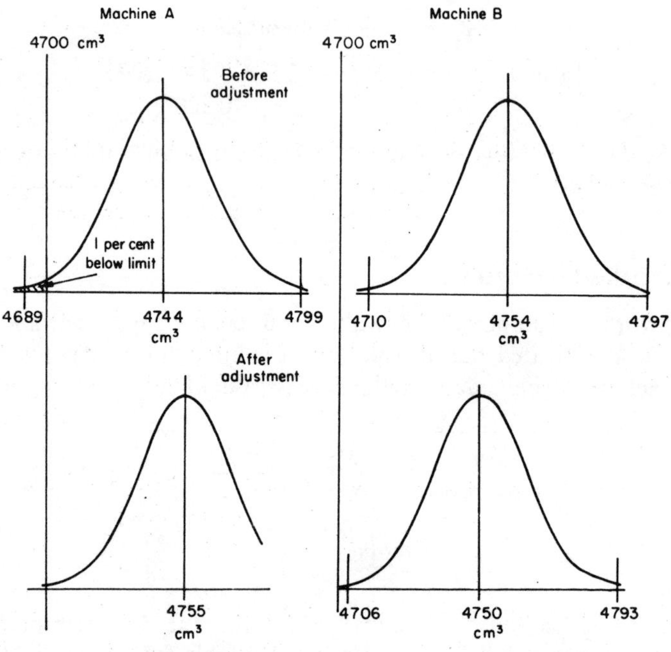

Fig. 6.10 Process capability (Data of section 6.6)

$$x = 4700 \quad \bar{x} = 4744 \quad s = 18{\cdot}4 \quad z = \frac{x-\bar{x}}{s} = \frac{-44}{18{\cdot}4} = -2{\cdot}39$$

From Table C, area to left of $z = 2{\cdot}39$ is $0{\cdot}5 - 0{\cdot}4913 = 0{\cdot}0087$

so about 1 per cent of the bottles filled by machine A contain less than the warranted minimum.

To reduce this to about 0·1 per cent it will be necessary to increase the average 'fill' by about 11 cm^3.

After adjusting both machines we can expect to get:

Machine	Mean	3 × Standard Deviation	Process capability limits cm³
A	4755	55·2	4699·8 to 4810·2
B	4750	43·5	4706·5 to 4793·5

Thus, in order to meet the warranted minimum, we have to accept that some bottles filled by machine A will contain as much as 4810 cm^3, while machine B seldom puts more than 4793 cm^3 into any bottle. Because it is less consistent, machine A is less efficient in that it will use more filling material, and hence be more costly.

N.B. Machine B could have its average 'fill' reduced by a further 6 cm^3 without running any real danger of producing a bottle containing less than 4·70 litres; however, the mean is not reduced below 4750 cm^3 in order to keep faith with the statement 'average contents 4·75 litres'.

6.7 Problems

1. Determine the probability that a sample of 8 pieces, drawn at random from a batch of 60 pieces containing 5 per cent defectives, will be found to include exactly 2 defectives.

   Determine also the mean and variance of the number of defectives expected to be found in samples of size 8 drawn, at random, from each of a series of batches similar to the one described above.
2. Samples of 20 pieces each are drawn, at random, from very large batches of components.
   (a) Use Table A to determine the probability that such a sample will contain:
      (i) 3 or more defectives
      (ii) Exactly 3 defectives
      (iii) Less than 3 defectives
      } when the batch contains 15 per cent defectives.

      (iv) Less than 2 defectives
      (v) Exactly 2 defectives
      (vi) More than 2 defectives
      } when the batch contains 3 per cent defectives.
   (b) Use Table B to obtain Poisson approximations to the probabilities asked for in (iv), (v), and (vi) above.
   (c) Determine the mean and variance of the number of defectives expected in samples of 20, drawn at random from very large batches containing

(i) 15 per cent,
(ii) 3 per cent defectives

for both the Binomial distribution and for an assumed Poisson approximation.

3. A certain firm uses a large fleet of delivery vehicles. Their records over a long period of time (during which their fleet size and utilization may be assumed to have remained sensibly constant) show that the average number of unserviceable vehicles per day is 3. Estimate the probability that, on a given day,
   (a) All their vehicles will be serviceable,
   (b) More than 2 vehicles will be unserviceable,
   (c) Exactly 4 vehicles will be unserviceable.
4. In setting up a Control Chart for Sample Means for a particular machining operation producing a nominal diameter of 0·450 in., 20 samples, each of 4 pieces, were taken at half-hourly intervals. It was found that the standard deviation of the means of the 20 samples was 0·0025 in. and the mean of the sample means was 0·4504 in.

   Determine
   (a) The probability that a given future sample will show a mean diameter
      (i) Greater than 0·4510 in.,
      (ii) Less than 0·4490 in.
   (b) The value of sample mean diameter which is likely to be exceeded
      (i) By only one in 40 samples,
      (ii) By only one in a thousand samples.
5. If 1000 very large batches of components, each containing 4 per cent defectives, have random samples each of 200 pieces drawn from them, how many of these samples would you expect to contain
   (a) More than 10 defectives,
   (b) Less than 5 defectives,
   (c) Exactly 2 defectives.
6. Discs stamped out of sheet metal are found to have a weight variation conforming to a Normal distribution with mean weight 0·2827 lb and standard deviation of weight 0·0003 lb. The specification weight for these discs is $0{\cdot}283 \pm 0{\cdot}0005$ lb.

   Determine
   (a) The percentage of discs which must be expected to be 'not to specification',
   (b) The minimum percentage 'not to specification' which could—theoretically—be achieved by adjusting the 'setting' to alter the distribution mean.

7. Control charts

7.1 Introduction

In any automatic, or semi-automatic, process there will be minor variations in the dimension produced from piece to piece, even if the 'setting' remains exactly the same. The object of Control Charting is to detect, as reliably as possible, when a change in dimension has occurred due to an 'assignable cause' (such as wear at the tool point or a shift in the setting), while allowing for the random piece-to-piece variation which is bound to occur.

It must be clearly understood from the start that a Control Chart is *not* a magic wand. It is quite ridiculous to suggest that mere use of a Control Chart can turn a basically poor 'process' into a good one.

Control charting is intended to provide a means of ensuring that a process remains 'in control', that is to say that the nature of the product has not changed since the start of the run. Further, the charts give warning of impending departure from permitted tolerances thus eliminating, or at least minimizing, the amount of unacceptable work produced. Properly designed, the charts should achieve their objective with a minimum amount of checking. In many cases they regularize checking which has previously been done on a haphazard basis, and may even reduce it in quantity.

Control Charts can only control a process within the 'capability' of that process. If this 'capability' is too wide to meet design tolerances the production of work outside limits is inevitable. This situation can only be rectified by a close study of the process to determine the cause, or causes, of variability in order either to eliminate them or at least reduce their effect. Alternatively, a review of the design tolerances may show them to be unnecessarily stringent.

Just as there are various types of 'process' in industry so also there are various types of Control Chart, each with its own advantages—and disadvantages—and with its own proper sphere of operation. It is important, if Control Charts are to produce the 'dividends' of which they are capable, that the right type of chart be used to control any particular process and that it be realized that certain 'processes', by their very nature, do not lend themselves to *statistical* Control Charting, e.g., 'processes' which are in no way 'automatic', the operator not working to a pre-determined stop or index-mark. Effectively each piece is 'one-off' and the operator directly controls the dimension produced on every piece, continually adjusting to keep to drawing.

When statistical charting in its more sophisticated forms is not applicable there may, in some cases, be something gained by 'incentive' charting, e.g., a chart showing the percentage of defectives previously achieved as a target to be beaten this time. Such charting would seem unlikely to be of any help when the operator is a skilled man—pride in his job provides more than adequate incentive—but for unskilled operators it could be of considerable value.

Over the course of this and the next two chapters we will consider a number of types of Control Charts, viz.,

(a) Pre-Control Charts.
(b) Control Charts for sample mean and range.
(c) Control Charts for proportion defective.
(d) Control Charts for number of defects per unit.

The latter three are statistically better founded but the system of Pre-Control Charting has the advantage of simplicity in use. Their advantages and limitations will be explained more fully in due course but, for the moment, a broad sub-division would be to say that Pre-Control Charts are applicable to semi-automatic processes, where the operator has the facilities to check each piece if necessary, and the charts would be plotted by the operator himself. Control Charts for sample mean and range are for controlling fully automatic processes, where the operator is probably responsible for three or more machines, and the charts might be plotted by a patrol inspector, for example.

7.2 Basic Pre-Control Chart technique

In its simplest form this system is based on the assumption that design tolerance and process capability are closely matched. If this is so then the width of the tolerance band is equal to six times the process standard deviation (σ). Assuming the dimension to be controlled is Normally distributed, this means that only 1 piece in 14 would be expected to lie in either one or other of the outer quarters of the tolerance band, and that only once in 196 (say 1 in 200) times would two consecutive pieces be expected to fall in one or other of these outer regions. This, when it happens, is taken as an indication that a shift in the setting of the machine is a more likely explanation, and calls for the process to be stopped and re-set.

Operation of a Pre-Control Chart would follow this sequence:

(a) Construct a chart consisting of four parallel lines representing the upper and lower tolerances and a pair of lines inside these each spaced one-quarter of the band width inside the tolerance lines. These inner lines are called Pre-Control Limits (P.C.L.). A chart of this nature must be prepared for each dimension to be controlled. See Fig. 7.1.

(b) When the process, or operation, is started take each piece as it is produced, measure and plot each controlled dimension for each piece. Continue until 5 consecutive pieces yield dimensions which are within the P.C.L.

(c) Start frequency gauging, i.e., take (say) every 20th piece produced, measure and plot all controlled dimensions. If these fall between P.C.L. continue frequency gauging.

(d) If a frequency gauged point falls outside P.C.L. (but inside design tolerance), measure and plot the dimensions of the very next piece produced; if these fall back within the P.C.L. revert to frequency gauging.

(e) If the two consecutive points both fall outside the P.C.L. stop the process and reset in respect of the offending dimension, or dimensions.

(f) After resetting start again at (b).

(g) If any single point falls outside the design tolerance, stop the process, rectify and start again at (b).

(h) If consecutive points fall outside *opposite* P.C.L.s, stop the process. This suggests excessive variability and an investigation must be carried out to establish the cause. After taking corrective action, start again at (b).

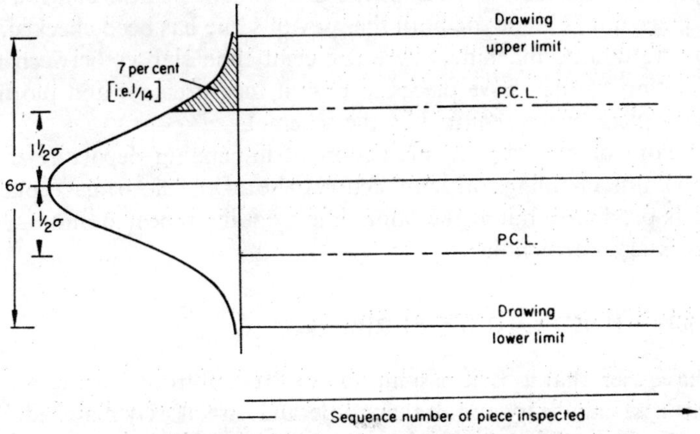

Fig. 7.1 Basic Pre-Control Chart

When the process capability is inadequate to meet the design tolerance it will be found difficult (or impossible) to obtain 5 consecutive pieces within P.C.L., thus precluding frequency gauging and maintaining maximum possible control by checking every piece. Thus Pre-Control Charting is a 'fail safe' system.

The choice of frequency for frequency gauging is initially made on a hit-or-miss basis: the frequency to aim for is such that about 25 (say 21 to 29) checks are made before resetting is called for. Consequently the frequency should be adjusted as the run proceeds, in order to converge on this target. With experience in the use of pre-control, and a knowledge of the production process concerned it seems likely that the initial 'guessed' frequency would be quite close to the optimum.

When a piece is being controlled on more than one dimension it is quite possible that the process capability may be adequate in respect of certain dimensions but not in respect of others. In this case, although every piece must be checked, only offending dimensions need to be measured every time, the others being checked on a frequency basis.

It will be evident, from the foregoing, that pre-control has the advantage that the charts can, and indeed must, be prepared beforehand by reference only to the drawing. Further it is arithmetically simple to operate and could be adapted for use by the machine operator even without charting, although this is felt to be somewhat unreliable, particularly when checking multiple dimensions. Further, charting provides both the operator with visual proof of the quality of his work, and the quality engineer with data relating to process capability, and evidence on which to base schemes for related processes at a later date.

By its very nature, pre-control can only be used on processes where the pieces are made one at a time, either under operator control, the next piece not being made until the previous one has been checked, or on an automatic machine where sufficient time elapses between the completion of successive pieces to permit measurement and plotting of every piece (when required by the scheme).

Pre-control provides an indication of impending departure from design limits, enabling corrective action to be taken before unacceptable work is produced, but at the same time avoiding repeated minor, and unnecessary, corrections.

7.3 Modified Pre-Control Charts

We have seen that a basic assumption in Pre-Control Charting is that the process capability and drawing tolerance are closely matched. We have also seen that if the process capability is inadequate to meet the drawing requirements ($6\sigma >$ drawing tolerance) the system will 'fail safe' and checking of every piece will continue. If this does happen it seems wise to carry out a process capability assessment to determine the limits which the process *is* capable of consistently meeting. The operator should then be warned against 'chasing his tail' (correcting for pieces produced 'not to drawing' when the error is, in fact, due to random variation) since this is bound to increase the percentage of defective work produced. Perhaps, in such circumstances, Pre-Control Charts using upper and lower limits based on 'process capability' instead of 'drawing' would be more realistic and of more value?

In the opposite case, i.e., process more than capable of meeting drawing requirements ($6\sigma <$ drawing tolerance) the system of Pre-Control, used in its basic form, would be 'safe' but would result in premature stopping for re-setting. This can be avoided if process standard deviation (σ) is established and a 'modified' Pre-Control Chart drawn up. Pre-Control lines on a modified' chart would be drawn at $1\frac{1}{2}\sigma$ inside the drawing upper and lower limits (instead of $\frac{1}{4}$ band width). If the 'drift' due to assignable causes—e.g., tool wear—is from undersize to oversize, then another line would be drawn at $4\frac{1}{2}\sigma$ above the drawing lower limit. The band from $1\frac{1}{2}\sigma$ to $4\frac{1}{2}\sigma$ above the drawing lower limit is the 'aiming band' into which the first 5 pieces made should fall, before frequency gauging is allowed. See Fig. 7.2.

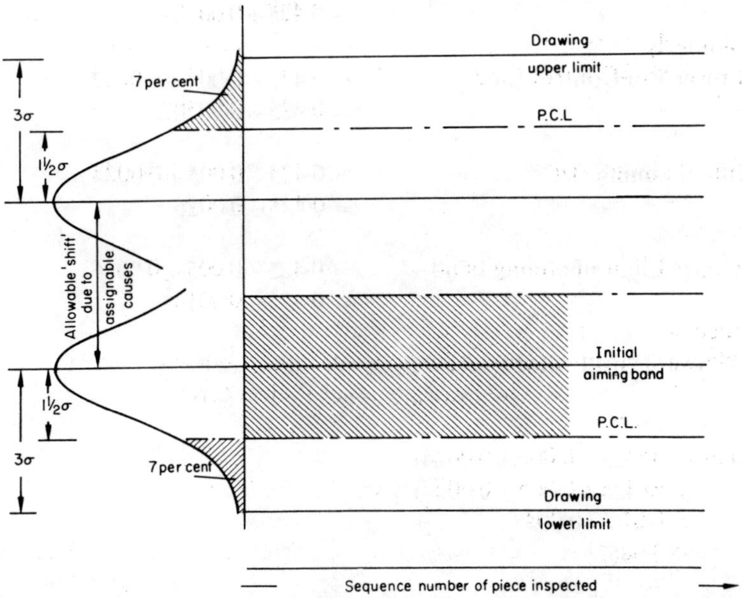

Fig. 7.2 Modified Pre-Control Chart

(If the 'drift' is such that size *reduces* progressively, then the aiming band would be similar but above the centre of the chart, instead of below.)

By using the Chart in this way the maximum safe 'drift' due to assignable causes is achieved and re-setting reduced to a minimum.

7.4 Modified Pre-Control Charts—worked example

On a broaching operation a dimension of $0{\cdot}425 \pm 0{\cdot}005$ in. is being produced. An investigation (see sections 7·7 and 7·8) has shown that the process standard deviation (σ) is 0·0008 in. and that a progressive increase of 0·001 in. per 50 pieces made is caused by the broaches wearing. Determine modified Pre-Control Limits suitable for this task and estimate the number of pieces likely to be made between re-sets. All dimensions are in inches:

Drawing Upper Limit $= 0{\cdot}425 + 0{\cdot}005$

$$\sigma = 0{\cdot}0008$$

$$1\tfrac{1}{2}\sigma = 0{\cdot}0012$$

Therefore Upper Pre-Control Line $= 0{\cdot}425 + 0{\cdot}005 - 0{\cdot}0012$
$= 0{\cdot}425 + 0{\cdot}0038.$

Similarly,
Lower Pre-Control Line $= 0{\cdot}425 - 0{\cdot}005 + 0{\cdot}0012$
$= 0{\cdot}425 - 0{\cdot}0038$

Initial aiming size $= 0{\cdot}425 - 0{\cdot}005 + 0{\cdot}0024$
$= 0{\cdot}425 - 0{\cdot}0026$

Upper Limit of aiming band $= 0{\cdot}425 - 0{\cdot}005 + 0{\cdot}0036$
$= 0{\cdot}425 - 0{\cdot}0014$

Also,
Allowable drift is from (drawing lower limit $+ 3\sigma$)
to (drawing upper limit $- 3\sigma$)

that is, $(0{\cdot}425 - 0{\cdot}005 + 0{\cdot}0024)$
to $(0{\cdot}425 + 0{\cdot}005 - 0{\cdot}0024)$
$= 0{\cdot}010 - 0{\cdot}0048$
$= 0{\cdot}0052.$

Now drift $= 0{\cdot}001$ per 50 pieces,

therefore probable number of pieces between re-sets

$$= \frac{0{\cdot}0052}{0{\cdot}001} \times 50$$

$$= 5{\cdot}2 \times 50 = 260 \text{ pieces.}$$

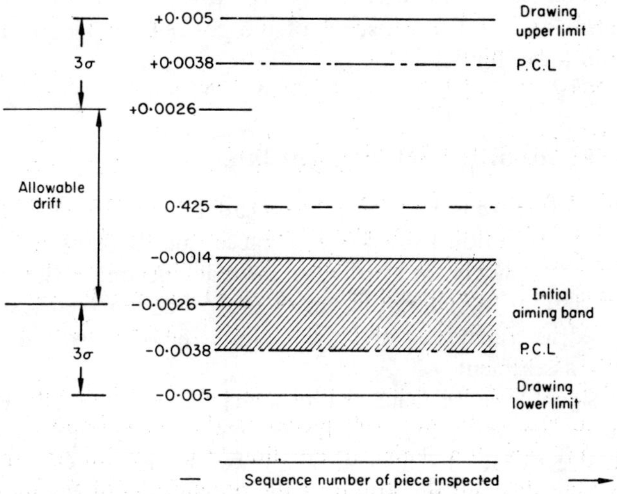

Fig. 7.3 Modified Pre-Control Chart (Data of section 7.4)

7.5 What if the assumptions are not valid?

The first assumption in the theory of Pre-Control Charting is that the dimensions produced are Normally distributed. If, in fact, the distribution is other than Normal the chance of a single piece being outside the P.C.L. is something other than 1 in 14, and hence the 1 in 200 assumption for the chance of 2 consecutive pieces falling outside the P.C.L. is incorrect. This means that, if the distribution is not Normal, the charts will still work but the 'risk' of making a decision to adjust unnecessarily will be different from the 1 in 200 we have assumed.

If the 'process capability' and 'drawing tolerance' are not closely matched (the second assumption is that they are) then either (a) checking of every piece continues when the process capability is not good enough—as already discussed in section 7.3—and the system 'fails safe'; or (b) when the process capability is more than adequate the basic system of Pre-Control would lead to premature adjustment—again 'safe'—but modified charts can be prepared to make the most efficient use of this desirable state of affairs.

7.6 Gauging

Since Pre-Control Charts do not depend upon the plotting of exact sizes, but only require the operator to know into which 'band' the dimension falls, it is possible to use limit gauges, etc. The only extra equipment needed is two gauges instead of one, since there are, in pre-control, four significant levels (drawing upper and lower limits and upper and lower P.C.L.) instead of the conventional two (drawing upper and lower limits).

7.7 Least squares best straight line

Whenever reference is made to 'process capability' we are dealing with the random variation from piece to piece due to many and varied causes, such as play in the bearings of the machine, uneven lubrication, minor variations in material hardness, etc. All assignable causes must be allowed for separately and their effects removed from a process capability assessment.

An 'assignable cause' (such as tool wear) may produce a progressive increase in size over, say, 200 pieces made—superimposed on the random size variation. This can be allowed for by drawing the 'best straight line' through the graph of the dimensions of the individual pieces, plotted in the sequence in which the pieces were produced (see the notes at the end of the chapter). This straight line then represents an estimate of the progressive movement of the *mean* of the distribution as production proceeds. Standard deviation (σ) for the process is then calculated using the deviation of the dimension of each piece from the appropriate mean (the later the piece is made the larger is the mean value appropriate to that piece).

7.8 Least squares best straight line—example

In practice a large number of pieces would be used but, for clarity, we will only use a few.

Production Sequence Number of Piece (x)	Dimension in $\times 0{\cdot}0001$ difference from Nominal 0·3750 in (y)
1	-6
21	-4
41	$+2$
61	$+3$
81	$+7$
101	$+11$

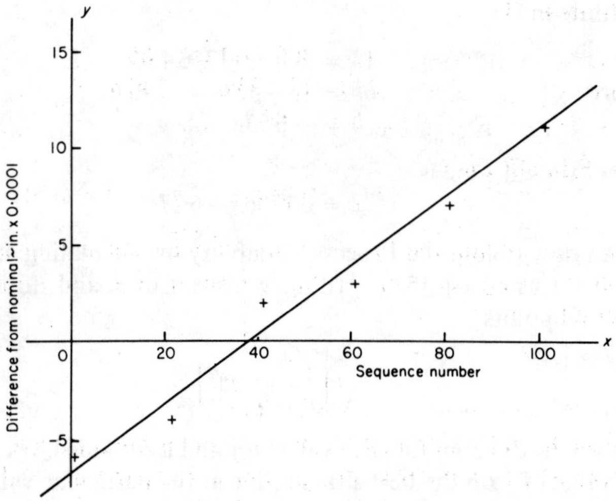

Fig. 7.4 'Least squares' best straight line (Data of section 7.8)

If the equation of the 'best straight line' is $y = ax+b$ then a and b are determined by solving the simultaneous equations:

$$\left.\begin{aligned}\Sigma y &= a\Sigma x + nb\\ \Sigma xy &= a\Sigma x^2 + b\Sigma x\end{aligned}\right\} \tag{7.1}$$

(See notes at end of the chapter.)

x	y	x^2	xy	
1	−6	1		−6
21	−4	441		−84
41	+2	1681	82	
61	+3	3721	183	
81	+7	6561	567	
101	+11	10201	1111	
306	+13	22606	1943 −90 = 1853	

$$n = 6$$
$$\Sigma x = 306$$
$$\Sigma y = 13$$
$$\Sigma x^2 = 22606$$
$$\Sigma xy = 1853$$

Putting these values in eq. (7.1):

$$\left.\begin{aligned}13 &= 306a+6b \\ 1853 &= 22606a+306b\end{aligned}\right\} \begin{matrix}(1)\\(2)\end{matrix}$$

$$(1)\times 51: \quad 663 = 15806a+306b \quad (3)$$

$$(2)-(3): \quad 1190 = 6800a$$

Therefore $$a = \frac{1190}{6800} = 0{\cdot}1756.$$

[This indicates a 'drift' of $0{\cdot}1756 \times 0{\cdot}0001$ in. per piece, i.e., $1{\cdot}756 \times 0{\cdot}001$ in. per 100 pieces made.]

Substitute in (1)

$$13 = 306 \times 0{\cdot}1756 + 6b$$

Therefore

$$6b = 13 - 53{\cdot}6 = -40{\cdot}6$$
$$b = -6{\cdot}77.$$

The Best Straight Line is

$$y = 0{\cdot}1756x - 6{\cdot}77$$

We can now obtain the Process Capability by calculating standard deviation (σ) using eq. (5.6), writing y instead of x and noting that $f = 1$ for all points:

$$\sigma = \sqrt{\left[\frac{1}{n}\Sigma(y-\bar{y})^2\right]}$$

where $\bar{y}$ will be different for each value of y and is obtained by calculating the value of y on the best straight line at the particular value of x. That is, $\bar{y} = 0{\cdot}1756x - 6{\cdot}77.$

x	$\bar{y}$	y	$y-\bar{y}$	$(y-\bar{y})^2$
1	−6·5944	− 6	+0·5944	0·3533
21	−3·0824	− 4	−0·9176	0·8420
41	+0·4296	+ 2	+1·5704	2·4662
61	+3·9416	+ 3	−0·9416	0·8866
81	+7·4536	+ 7	−0·4536	0·2057
101	+10·9656	+11	+0·0344	0·0012
				4·7550

$$n = 6 \quad \Sigma(y-\bar{y})^2 = 4{\cdot}7550$$

therefore

$$\sigma = \sqrt{0{\cdot}7925}$$
$$= 0{\cdot}8902 \times 0{\cdot}0001 \text{ in.}$$

Process capability $= \pm 3\sigma = \pm 2{\cdot}67 \times \frac{1}{10000}$ in.

$= \pm 0{\cdot}3 \times 0{\cdot}001$ in., say.

And drift $= 1{\cdot}8 \times 0{\cdot}001$ in. per 100 pieces (approx.).

MATHEMATICAL NOTES

7.9 Derivation of least squares best straight line formulae

Let $y = \alpha x + \beta$ be the exact, but unknown, relationship between the n pairs of values $(x_i y_i)$. $(i = 1, 2, 3, 4, \ldots n)$.

Let $y = ax + b$ be the 'least squares' best estimate of the true relationship $y = \alpha x + \beta$, i.e., $y = ax + b$ is a straight line drawn through the graph plotted for all the points $(x_i y_i)$ where a and b are such that the sum of the squares of the distances of the points from the line (measured in the y-direction) is minimized.

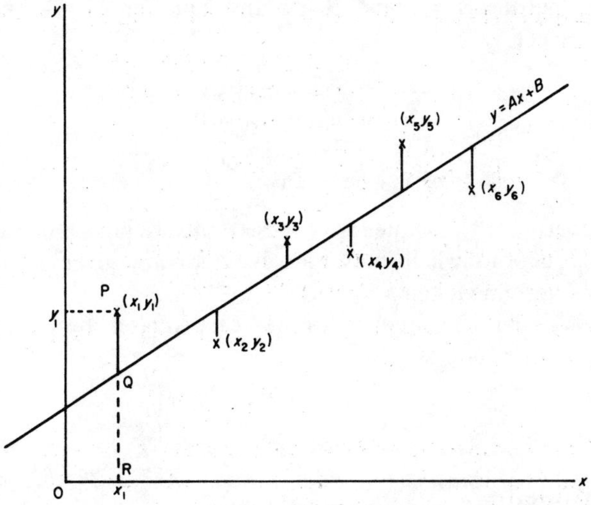

Fig. 7.5 'Least squares' best straight line

Let $y = Ax+B$ be *any* straight line drawn on the graph. Now, referring to Fig. 7.5

$$\begin{aligned} PR &= y_1 \\ QR &= Ax_1+B. \\ PQ &= PR-QR \\ &= y_1-(Ax_1+B). \end{aligned}$$

For the *best* straight line the sum of the squares of all such distances PQ (for all the points) is to be a minimum. Let V denote this sum of squares.

$$\begin{aligned} V &= [y_1-(Ax_1+B)]^2+[y_2-(Ax_2+B)]^2+\text{etc.} \\ &= \sum_i [y_i^2-2y_i(Ax_i+B)+(Ax_i+B)^2] \\ &= \sum_i y_i^2-2A\Sigma x_i y_i-2B\Sigma y_i+A^2\Sigma x_i^2+2AB\Sigma x_i+B^2 n. \end{aligned}$$

We have to find particular values (a and b) for A and B such that V is a minimum, that is to say, when

$$A = a \text{ and } B = b, \frac{\partial V}{\partial A} = 0 \text{ and } \frac{\partial V}{\partial B} = 0.$$

Now

$$\frac{\partial V}{\partial A} = -2\Sigma x_i y_i+2A\Sigma x_i^2+2B\Sigma x_i$$

$$\frac{\partial V}{\partial B} = -2\Sigma y_i+2A\Sigma x_i+2Bn.$$

Hence, putting $A = a$ and $B = b$ and equating $\partial V/\partial A$ and $\partial V/\partial B$ to zero, we get

$$\left.\begin{aligned} 2a\Sigma x_i^2 + 2b\Sigma x_i - 2\Sigma x_i y_i &= 0 \\ 2a\Sigma x_i + 2bn - 2\Sigma y_i &= 0 \end{aligned}\right\}$$

therefore

$$\left.\begin{aligned} a\Sigma x_i^2 + b\Sigma x_i &= \Sigma x_i y_i \\ a\Sigma x_i + bn &= \Sigma y_i \end{aligned}\right\}$$

If we rearrange the sequency of these terms we have a pair of simultaneous equations which may be solved, for any given set of points $(x_i y_i)$, to determine a and b.

If $y = ax + b$ is the equation of the 'least squares' best straight line for a given set of points then

$$\left.\begin{aligned} \Sigma y &= a\Sigma x + nb \\ \Sigma xy &= \Sigma x^2 + b\Sigma x \end{aligned}\right\} \qquad (7.1)$$

7.10 Problems

1. In a certain component a bore is being produced to a nominal diameter of 0·348 ± 0·006 in. Construct a basic Pre-Control Chart for this dimension.

   The following table shows the measured bore of a number of pieces, as produced. Plot these on the Chart and write a brief commentary.

Sequence number of piece	1	2	3	4	5	10	15
Bore (in.)	0·347	0·347	0·349	0·348	0·350	0·350	0·346
	20	25	30	31	35	40	45
	0·348	0·348	0·352	0·350	0·350	0·351	0·346
	50	55	56	57	58	59	60
	0·346	0·344	0·343	0·347	0·350	0·350	0·344
	61	62	63	64	65	70	75
	0·346	0·346	0·348	0·349	0·350	0·350	0·347
	80	85	90	95	100	101	102
	0·348	0·348	0·347	0·349	0·344	0·352	0·347
	103	104	105	106	111	116	121
	0·347	0·348	0·346	0·349	0·347	0·346	0·348
	126	131	136	141			
	0·348	0·347	0·349	0·340			

2. A slot is to be cut to a nominal depth of 0·465 ± 0·010 in. An earlier process capability assessment has shown that the standard deviation of depth for this operation is 0·002 in. and that there is also a progressive decrease in actual depth of cut of about 0·001 in. for every

60 pieces machined, due to tool wear. Construct a suitable Pre-Control Chart and estimate the number of pieces you would expect to be machined before adjustment to depth of cut becomes necessary.

3. A grinding operation was subjected to an investigation to establish to what limits it could consistently produce and also the progressive rate of change of dimension caused by grinding wheel wear. Fifty pieces were produced, without adjustment, and their diameters, in the same sequence as made, were measured to 0·0001 in. and are recorded below. Calculate the approximate number of pieces made per 0·001 in. change in diameter, and the process standard deviation.

Sequence number of piece	20	40	60	80	100
Diameter (in.)	0·2476	0·2490	0·2500	0·2504	0·2520
	120	140	160	180	200
	0·2524	0·2530	0·2542	0·2542	0·2548

The drawing for this operation calls for a dimension of 0·2500 ±0·0020 in.; construct a suitable modified Pre-Control Chart for the operation, and estimate the number of pieces expected to be produced between successive adjustments (due to wheel wear).

Reference

Juran, J. M. (Editor), *Quality Control Handbook* [Ch. 19, Pre-Control]. McGraw-Hill, 1951.

8. Control charts for sample mean and range

8.1 Properties of the distribution of sample means

If a large number (N) of random samples, each of n pieces, is taken from a population whose mean is μ and variance σ^2, and the mean of each sample is determined [$\bar{x}_i$ denoting the mean of the ith sample; ($i = 1, 2, 3 \ldots N$)], then the distribution of these sample means can be formed.

Let $\bar{X}$ be the mean, and s^2 the variance of the distribution of sample means, that is

$$\bar{X} = \frac{1}{N}\sum_i \bar{x}_i \qquad (8.1)$$

$$s^2 = \frac{1}{N}\sum_i (\bar{x}_i - \bar{X})^2 \qquad (8.2)$$

Then it can be shown that

$$\bar{X} = \mu \qquad (8.3)$$

That is, *the mean of the distribution of sample means is equal to the population mean*, and

$$s^2 = \frac{\sigma^2}{n} \qquad (8.4)$$

Variance of distribution of the means of samples of n pieces is equal to the population variance divided by n.

Further, even if the parent population from which the samples are taken is *not* Normally distributed, the distribution of sample means gets nearer to the Normal as the sample size gets larger: means of samples of 8 pieces will have a distribution which is more nearly Normal than the distribution of means of samples of 4 pieces from the same non-Normal population.

N.B. Even for 'U' shaped or 'reverse J' shaped distributions, the distribution of sample means is sensibly Normal for samples of 8 or more pieces.

In the following paragraphs the system of symbols and notation set out in B.S. 2564: 1955 will be used.

8.2 **Sample range estimate of standard deviation**

It will be readily obvious that there is a connection between the average range ($\bar{w}$) of samples of a given size, and the standard deviation (σ) of the population from which they are drawn. The theory of the relationship is too complex to be examined here, but it is sufficient to say that a factor d_n has been determined such that $\bar{w}$ divided by d_n gives a good estimate of σ.See Table G. (Para. 1). The advantage of obtaining σ in this way lies in the considerable reduction in the amount of calculation needed.

8.3 **Control Charts for mean and range**

These charts are based on process capability, which must first be established, either by taking say the first 50 or 100 pieces produced after the process has 'settled down' and computing the mean and standard deviation, or by taking say the first ten samples of the chosen size (perhaps 4), determining the mean and range for each sample, hence the mean of the means (process mean) and mean range, using the latter to estimate the process standard deviation.

Again assuming the sample mean of the controlled dimension to be Normally distributed—remember it is true to say that even if the distribution of the particular dimension for single pieces is *not* Normal, the distribution of sample means will be very close to Normal—a Control Chart for sample means is constructed by drawing a line mid-way between the design limits together with two pairs of parallel lines, above and below the central line.

These lines are so placed that random piece-to-piece variation will only cause the mean of one sample in forty to fall outside the inner pair, and of one sample in a thousand to fall outside the outer pair. Consequently, unless the setting of the machine changes, it is unlikely that a sample mean will fall outside the inner pair of lines, if one does it is taken as a *warning* that the setting may have altered and so the inner pair of lines are called 'warning limits'. The chance that random variation will produce a sample mean which is outside the outer pair of lines is so small that, when it happens, it is taken as a strong indication that the setting has probably changed and the machine is stopped and adjusted. Consequently the outer pair of lines are called 'action limits'. It must be realized that a sample mean outside the Action Limits does not *guarantee* that the setting has altered; this could happen due to random variation about the correct setting. However, this is so rare that a far more likely reason is a change due to an assignable cause.

Sample range is related to standard deviation and is used as an indication of any increase in process variability. Since the chart for means is based on an assumed standard deviation it is necessary to chart sample range as well as mean. A control chart for sample range is drawn up incorporating an Upper Control Limit, being the value of sample range which would be expected to be exceeded only once in forty times, unless the process variability increased.

A pair of charts (for mean and range) would be drawn up for each dimension to be controlled.

Samples of the specified number of pieces are taken at random from the batch at regular intervals and the appropriate dimensions are measured. The sample mean and range for each dimension are then calculated and plotted on the charts. If they fall inside the 'warning limits' the samples continue to be taken at the same intervals, if they fall between 'warning' and 'action' limits sampling should be intensified, and if a point falls outside the 'action' limits the process must be stopped and the appropriate dimension, or dimensions, reset.

If the sample range is above the Upper Control Limit this suggests an unacceptable increase in process variability, the process should be stopped, the cause found and rectified.

When the Control Charts are intended to give an indication of the current state of production the sample should, in each case, contain only pieces taken from the last few produced. However, if the charts are to give an overall quality assurance for the entire output of the machine, then each sample should be chosen at random from all the pieces made by the machine since the previous sample was taken.

8.4 **Control Charts for mean and range—worked example**

A machine manufactures plugs to a nominal diameter of $0{\cdot}500 \pm 0{\cdot}030$ in. Calculate control limits for the mean diameter for this machine from the following data. (10 samples of 5 plugs each.) Draw the control chart and plot the first 10 points. Also, calculate the 1 in 40 control limits for range and plot the range chart.

Sample	1	2	3	4	5	6	7	8	9	10
All	0·488	0·494	0·498	0·492	0·491	0·501	0·497	0·495	0·504	0·500
diameters	0·489	0·494	0·515	0·509	0·509	0·501	0·492	0·499	0·505	0·493
in	0·505	0·499	0·487	0·514	0·499	0·505	0·496	0·498	0·513	0·506
inches	0·494	0·489	0·505	0·491	0·509	0·482	0·503	0·497	0·493	0·493
	0·500	0·505	0·511	0·501	0·498	0·520	0·492	0·487	0·503	0·502
$\bar{x}$	0·4952	0·4962	0·5032	0·5014	0·5021	0·5018	0·4960	0·4952	0·5036	0·4988
Range w	0·017	0·016	0·028	0·023	0·018	0·038	0·011	0·012	0·020	0·013

Number of samples = 10
Sum of samples means = $\Sigma(\bar{x}) = 4{\cdot}9926$

Therefore process mean (mean of sample means) $= \dfrac{4{\cdot}9926}{10}$

i.e., $\bar{X} = 0{\cdot}49926$ in.

(This is approximately 0·0007 in. below the nominal 0·500 in. diameter but, since the tolerance is $\pm 0{\cdot}030$ in. it seems unlikely that 're-setting' would achieve a closer value.)

We will take $\bar{X} = 0{\cdot}500$ in. from now on.

Sum of sample ranges $= \Sigma w = 0{\cdot}196$

Therefore sample mean range $= \dfrac{0{\cdot}196}{10}$

that is, $\bar{w} = 0{\cdot}0196$

Sample size $n = 5$

From Table G, Para. 1: $d_n = 2{\cdot}33$.

$$\begin{aligned}\text{Estimate of } \sigma &= \bar{w} \div d_n \qquad (8.5)\\ &= 0{\cdot}0196 \div 2{\cdot}33\\ &= 0{\cdot}00843 \text{ in.}\end{aligned}$$

(Tolerance = $\pm 0{\cdot}03$ in., therefore process is theoretically capable of meeting drawing limits.)

Standard deviation of sample means

$$= \frac{\sigma}{\sqrt{n}}$$

$$= \frac{0{\cdot}008426}{\sqrt{5}} = 0{\cdot}003768 \text{ in.}$$

Sample means are Normally distributed, hence, if process mean remains unchanged there is only 1 chance in 40 of obtaining a sample mean greater than $\bar{X} + 1{\cdot}96\sigma/\sqrt{n}$, and only 1 in 40 of obtaining a sample mean less than $\bar{X} - 1{\cdot}96\sigma/\sqrt{n}$. If such a sample is obtained it is taken to be a warning that the process mean may have changed.

So the warning limits are $\bar{X} \pm 1{\cdot}96\sigma/\sqrt{n}$

$$= 0{\cdot}500 \pm 1{\cdot}96 \times 0{\cdot}003\,768$$
$$= 0{\cdot}500 \pm 0{\cdot}0074 \text{ in.}$$

Similarly there is only 1 chance in 1000 of obtaining a sample mean outside the band $\bar{X} \pm 3{\cdot}09\sigma/\sqrt{n}$. Consequently, if this does happen, it is taken to be a strong indication that the process mean has changed, and action is taken to 're-set'.

Thus the action limits are

$$\bar{X} \pm 3{\cdot}09\sigma/\sqrt{n}$$
$$= 0{\cdot}500 \pm 0{\cdot}0116 \text{ in.}$$

Since σ was obtained from $\bar{w} \div d_n$ we can write

$$\frac{1{\cdot}96\sigma}{\sqrt{n}} = \frac{1{\cdot}96\bar{w}}{d_n\sqrt{n}}$$

and

$$\frac{3{\cdot}09\sigma}{\sqrt{n}} = \frac{3{\cdot}09\bar{w}}{d_n\sqrt{n}}$$

Hence by tabulating two new constants (Table G, Para. 2)

$$A'_{0{\cdot}025} = \frac{1{\cdot}96}{d_n\sqrt{n}}$$

$$A'_{0{\cdot}001} = \frac{3{\cdot}09}{d_n\sqrt{n}}$$

we can obtain our warning and action limits more simply, thus:

$$\tfrac{1}{40}\text{ (Warning) limits: } \bar{X} \pm \bar{w}\,.\,A'_{0{\cdot}025} \qquad (8.6)$$

$$\tfrac{1}{1000}\text{ (Action) limits: } \bar{X} \pm \bar{w}A'_{0{\cdot}001} \qquad (8.7)$$

The control chart for sample means can now be drawn. (Fig. 8.1).

Since the control limits depend on σ, it is necessary to check that σ is not likely to have changed, consequently a control chart for sample range (w)—which depends on σ—is also required.

If the process variability (i.e., standard deviation) remains unchanged, then there is only 1 chance in 40 of getting a sample whose range exceeds $\bar{w} \times D'_{0{\cdot}975}$ and similarly 1 chance in 40 of getting a sample whose range is less than* $\bar{w} \times D'_{0{\cdot}025}$. The factors $D'_{0{\cdot}975}$ and $D'_{0{\cdot}025}$ are given in Table G. (Para. 3).

$$n = 5, \; D'_{0{\cdot}025} = 0{\cdot}37, \; D'_{0{\cdot}975} = 1{\cdot}81$$

*This is, of course, a desirable state of affairs so, if it happens, we would like to know why in order to perpetuate it and—if possible—extend the benefit to other similar machines.

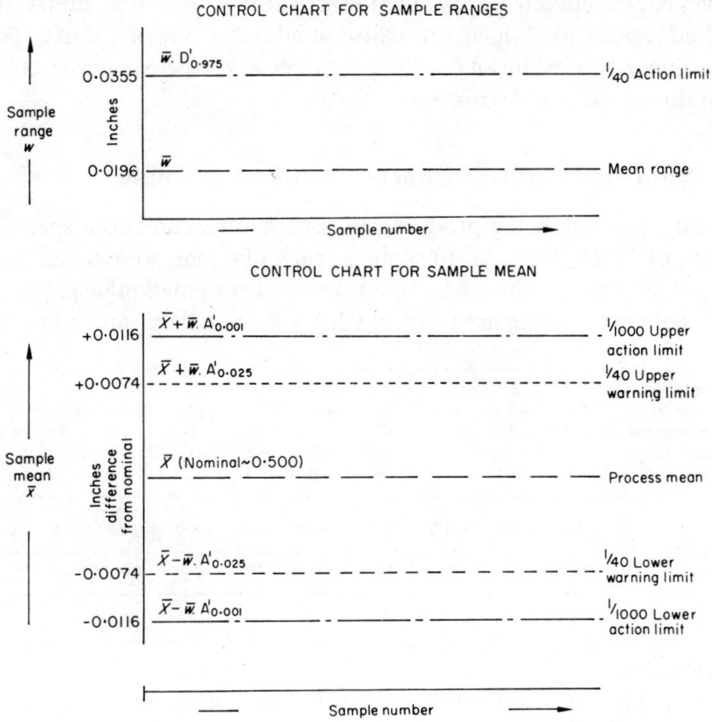

Fig. 8.1 Control Charts for sample mean and range

Limits for sample range:

Upper limit $= \bar{w} \times D'_{0\cdot975}$ (8.8)
$= 0{\cdot}0196 \times 1{\cdot}81$
$= 0{\cdot}0355$ in.

Lower limit $= \bar{w} \times D'_{0\cdot025}$ (8.9)
$= 0{\cdot}0196 \times 0{\cdot}37$
$= 0{\cdot}0073$ in.

The control chart for sample range can now be drawn (Fig. 8.1). The ‘lower limit’ has not been plotted since this is not an ‘action’ limit.

8.5 Modified Control Charts

When the process capability is more than adequate to meet the design tolerance the control on sample means can be relaxed, since the process mean can be allowed to vary to some extent before there is any real likelihood of producing pieces outside design limits. This allows a wider band between upper and lower warning or action limits, the revised values for which are called modified Control Limits. This variation in process mean may be very necessary if there is a progressive drift due to assignable causes.

8.6 Modified Control Charts—worked example

Locking pins are being produced by a certain machine to a specified length of 2·350±0·025 in. 10 samples, each of 4 pins, were taken from the output of the machine when it was known to be functioning properly. Use these data to prepare Control Charts for sample mean and range.

Sample No.	1	2	3	4	5	6	7	8	9	10
Difference	+4	+10	+6	0	+2	−1	+1	+3	+3	+7
from nominal	−2	+10	0	−1	0	−5	0	+1	−1	+2
(2·350 in.)	−3	+2	−1	−1	−2	−6	−6	0	−5	−2
in. × $\frac{1}{1000}$	−5	0	−3	−6	−9	−8	−7	−3	−7	−4
$\bar{x}$	−1·5	+5·5	+0·5	−2·0	−2·25	−5·0	−3·0	+0·25	−2·5	+0·75
w	9	10	9	6	11	7	8	6	10	11

10 samples.

$\Sigma\bar{x} = -9{\cdot}75 \quad \bar{X} = -0{\cdot}000\,975$ difference from nominal,

i.e., mean length $= 2{\cdot}350 - \dfrac{0{\cdot}975}{1000}$

$= 2{\cdot}349$ in.

$\Sigma w = 87 \qquad \bar{w} = 0{\cdot}0087$ in.

Sample size $n = 4$.

From Table G, Para. 1

$d_n = 2{\cdot}06$

Estimate of σ $= \bar{w} \div d_n$

$= 8{\cdot}7 \div 2{\cdot}06 \times 0{\cdot}001$ in.

$= 4{\cdot}226 \times 0{\cdot}001$ in.

Process Capability $= \pm 3\sigma = \pm 0{\cdot}0127$ in. (say 0·013 in.)

Tolerance $= \pm 0{\cdot}025$ in.

The process is more than capable of meeting the specification and the dimension mean may be allowed to vary from 0·013 in. above the drawing lower limit to 0·013 in. below the drawing upper limit (from 0·012 in. below nominal to 0·012 in. above; 0·024 in. allowable drift for assignable causes. To achieve this maximum safe 'drift' it would be necessary to re-set in order to bring the initial process mean length down by about 0·011 in.)

Drawing upper limit = $T_u = 2{\cdot}350+0{\cdot}025$
Upper limit for process mean = $\bar{X}_u = T_u - 3\sigma$
$= 2{\cdot}350+0{\cdot}012$
Drawing lower limit = $T_L = 2{\cdot}350-0{\cdot}025$
Lower limit for process mean = $\bar{X}_L = T_L + 3\sigma$
$= 2{\cdot}350-0{\cdot}012$

From Table G, Para. 2

$$n = 4 \qquad A'_{0\cdot025} = 0{\cdot}48 \qquad A'_{0\cdot001} = 0{\cdot}75$$

so $\frac{1}{40}$ Warning limits for sample mean are

$\bar{X}_u + \bar{w}.A'_{0\cdot025}$
$= 2{\cdot}350+0{\cdot}012+(0{\cdot}0087\times 0{\cdot}48)$
$= 2{\cdot}350+0{\cdot}0162$
and $\bar{X}_L - \bar{w}.A'_{0\cdot025}$
$= 2{\cdot}350-0{\cdot}0162.$

And $\frac{1}{1000}$ Action limits for sample mean are
$\bar{X}_u + \bar{w}.\ A'_{0\cdot001}$
$= 2{\cdot}350+0{\cdot}012+(0{\cdot}0087\times 0{\cdot}75)$
$= 2{\cdot}350+0{\cdot}0185$
and $\bar{X}_L - \bar{w}.A'_{0\cdot001}$
$= 2{\cdot}350-0{\cdot}0185.$

The modified Control Chart for sample mean can now be drawn (Fig. 8.2, page 102.)
Control limits for sample range are

$\frac{1}{40}$ Upper limit: $\bar{w}\times D'_{0\cdot975} = 0{\cdot}0087\times 1{\cdot}93 = 0{\cdot}0168$
$\frac{1}{40}$ Lower limit: $\bar{w}\times D'_{0\cdot025} = 0{\cdot}0087\times 0{\cdot}29 = 0{\cdot}0025$

8.7 **Relative precision index (R.P.I.)**

If the specification tolerance for a component is considerably less than 6σ for the process used to produce the component we know that the production of some pieces 'not to specification' is inevitable; the precision of the process is low compared with the precision required, i.e., of low relative precision. If the specification tolerance is about equal to 6σ then very few pieces 'not to specification' would be expected; the process is said to have medium relative precision. Thirdly, if the specification tolerance is considerably greater than 6σ then the process is said to have high relative precision and 'modified' Control Charts for sample mean can be used to maintain the quality of production. (See Fig. 8.3, page 103.)

As we saw in section 8.2, sample average range $(\bar{w})\div d_n$ gives a good estimate of σ. Consequently, for samples of a given number of pieces

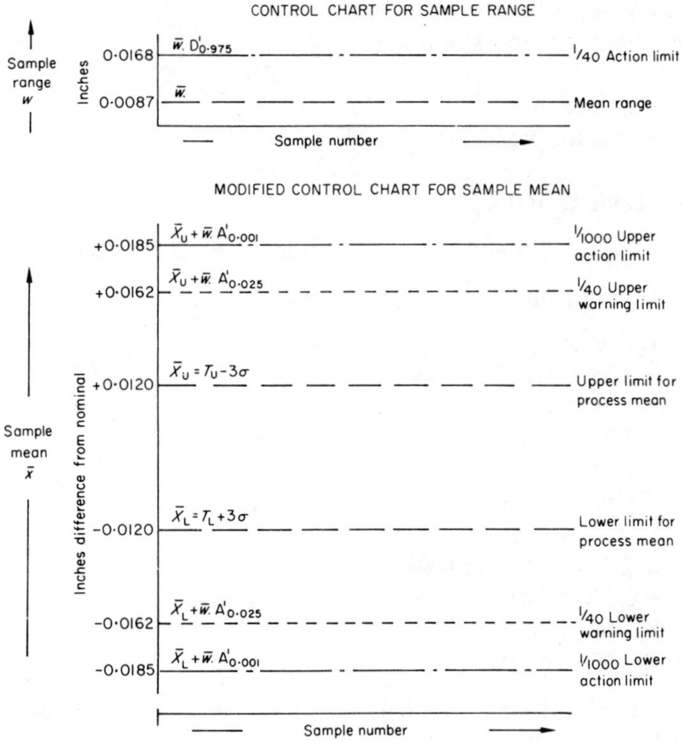

Fig. 8.2 Modified Control Charts for sample mean and range

(n), $\bar{w}$ is proportional to σ and specification tolerance divided by sample average range is a measure of relative precision, and is called the relative precision index (R.P.I.).

$$\text{Relative precision index (R.P.I.)} = \frac{\text{Specification tolerance}}{\text{Sample average range}}$$

The values of R.P.I., for various sample sizes (n), which indicate low, medium, and high relative precision are tabulated in Table G, Para. 4.

8.8 Problems

1. Draw control charts for mean and range based on the ten samples of 5 spacer lengths shown in table 8.1, page 104.

   What proportion of production may be expected to lie outside tolerance limits of $0{\cdot}420 \pm 0{\cdot}010$ in.? If it were possible to shift the process mean without changing the average range, to what value should the mean be moved in order that the smallest proportion

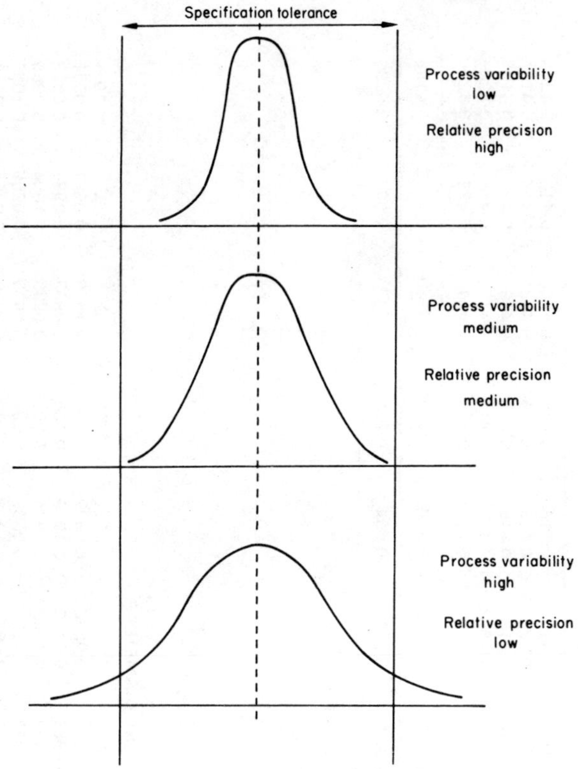

Fig. 8.3 Process variability and relative precision

fall out of the limits and what would this proportion be? (Assume that the lengths are Normally distributed.)

2. Calculate $\frac{1}{40}$ and $\frac{1}{1000}$ control limits for the mean value of the diameter of pins produced by a particular process using the data in table 8.2, page 104.

   Also determine the one in forty upper and lower limits for the sample range and plot the first ten points on each chart.

3. The drawing, relating to the pins measured and tabulated in question 2, calls for a diameter of $0{\cdot}100 \pm 0{\cdot}005$ in. Prepare a modified control chart for sample mean, allowing for the maximum 'safe' shift of process mean due to assignable causes.

4. The following are means and ranges of 20 samples of 6 bores each (in inches).

   Use the first 10 samples to set up Control Charts for sample mean and range to control future production.

   Plot the next 10 samples. What proportion of production may be

Table 8.1

SAMPLE	1	2	3	4	5	6	7	8	9	10
Length in.	0·411	0·414	0·414	0·419	0·419	0·414	0·419	0·412	0·414	0·421
	0·418	0·421	0·415	0·418	0·420	0·413	0·414	0·413	0·414	0·413
	0·414	0·419	0·415	0·426	0·416	0·416	0·419	0·419	0·422	0·416
	0·414	0·416	0·418	0·414	0·409	0·417	0·415	0·415	0·414	0·422
	0·418	0·418	0·416	0·416	0·416	0·416	0·419	0·414	0·422	0·422
Mean	0·4150	0·4176	0·4156	0·4186	0·4160	0·4152	0·4172	0·4146	0·4172	0·4188
Range	0·007	0·007	0·004	0·012	0·011	0·004	0·005	0·007	0·008	0·009

$\bar{X} = 0{\cdot}4166 \qquad \bar{w} = 0{\cdot}0074$ in.

Table 8.2

PIN DIAMETERS—inches

Sample	1	2	3	4	5	6	7	8	9	10
	0·1006	0·0998	0·1005	0·1010	0·0983	0·0987	0·1004	0·1008	0·0997	0·0981
	0·1011	0·1026	0·1012	0·0988	0·0980	0·1009	0·1001	0·0993	0·0987	0·0993
	0·0989	0·1019	0·1009	0·0992	0·1003	0·0985	0·1003	0·0999	0·1004	0·0995
	0·1000	0·1002	0·1019	0·0998	0·1004	0·0987	0·1011	0·0999	0·0999	0·1007
	0·0989	0·0998	0·1002	0·1000	0·0991	0·1002	0·1003	0·0999	0·0993	0·0986

expected to lie outside tolerance limits of 0·435 ± 0·005 in., assuming the bores to be Normally distributed.

Sample	Mean	Range	Sample	Mean	Range
1	0·4372	0·010	11	0·4370	0·006
2	0·4324	0·009	12	0·4322	0·004
3	0·4318	0·008	13	0·4356	0·010
4	0·4344	0·004	14	0·4322	0·005
5	0·4346	0·005	15	0·4306	0·008
6	0·4332	0·011	16	0·4352	0·011
7	0·4340	0·009	17	0·4332	0·006
8	0·4344	0·003	18	0·4346	0·005
9	0·4308	0·002	19	0·4360	0·004
10	0·4350	0·006	20	0·4374	0·006

5. 100 set screw blanks produced from hexagon bar by an automatic machine were each measured for 3 dimensions.

Nominal sizes:

Head thickness $0{\cdot}125\ ^{+0}_{-0{\cdot}010}$ in.

Overall length $0{\cdot}525\ ^{+0{\cdot}015}_{-0{\cdot}010}$ in.

Shank diameter $0{\cdot}186\ ^{+0}_{-0{\cdot}002}$ in.

The 100 measurements yielded the following distributions:

Head thickness (in.)	No. of blanks	Overall length (in.)	No. of blanks	Shank Diameter (in.)	No. of blanks
0·1200	1	0·527	4	0·1845	2
0·1220	1	0·528	21	0·1850	26
0·1225	8	0·529	34	0·1855	39
0·1230	13	0·530	22	0·1860	33
0·1235	7	0·531	16		
0·1240	45	0·532	3		
0·1245	12				
0·1250	9				
0·1255	2				
0·1260	1				
0·1280	1				

(a) Calculate the mean and standard deviation of each of the three dimensions.

(b) Comment on any adjustment to setting which you think is necessary, on the basis of the above data.

(c) On the assumption that setting has been adjusted, where necessary, to put process mean at mid-tolerance, set up Control Charts for mean and range of samples of 4 pieces, to control future output.

Reference:

B.S. 2564: 1955, B. P. Dudding and W. J. Jennett, *Control Chart Technique when Manufacturing to a Specification*, B.S.I.

9. Control charts for proportion defective and for defects per unit (p charts and d charts)

9.1 Introduction

Control charts for sample mean and range require that the product being controlled shall be assessed quantitatively (e.g., diameter of a shaft, length of a pin, depth of a slot). For products which are subject to a qualitative assessment—'good' or 'bad'—(e.g., explosives, which either fire or they don't; soldered joints, which either make electrical contact or they don't; dimensions checked on go/no go gauges) another form of charting is required. Control in such cases is achieved on the basis of the *proportion* of pieces found to be defective—p charts.

Similarly there are products where a number of defects per piece may be inevitable, and acceptable if controlled below a certain limit (e.g., blowholes in castings, air bubbles in glass, end breaks in lengths of yarn). In such products control is maintained on the basis of the *number* of defects per piece—d charts.

9.2 p-charts

If the mean proportion defective ($\bar{p}$) remains constant, then the proportion of defectives (p) in each sample will vary from sample to sample and will form a Binomial distribution having the following properties:

$$\text{Mean proportion defective} = \bar{p}$$

$$\text{Standard deviation of proportion defective} = \sqrt{\frac{\bar{p}\bar{q}}{n}}$$

(These were obtained from eq. (6.5) and square root of eq. (6.6), by dividing by n to change from *number* defective to *proportion* defective.)

where $\bar{p}$ = mean of the proportions defective found in the samples used to set up the chart.

$\bar{q} = 1 - \bar{p}$

n = sample size.

We know that, for a Normal distribution, only one member in a thousand exceeds the mean by more than 3·09 standard deviations; now if n is reasonably large (see section 6.6) the Normal is a good approximation to the Binomial and hence there is only one chance in a thousand of getting a sample whose proportion defective, p, exceeds

$$\bar{p} + 3{\cdot}09\sqrt{\frac{\bar{p}\bar{q}}{n}}$$

Consequently this value is taken as the 'action' limit for proportion defective. (In practice it is common to use $3\sqrt{(\bar{p}\bar{q}/n)}$ instead of $3{\cdot}09\sqrt{(\bar{p}\bar{q}/n)}$.

Similarly there is only 1 chance in 40 of getting a sample whose proportion defective, p, exceeds

$$\bar{p} + 1{\cdot}96\sqrt{\frac{\bar{p}\bar{q}}{n}}$$

This value is, therefore, used as a 'warning' limit for proportion defective. (Again it is common practice to use $2\sqrt{(\bar{p}\bar{q}/n)}$ instead of $1{\cdot}96\sqrt{(\bar{p}\bar{q}/n)}$.)

If the population proportion defective has not changed there is only 1 chance in 1000 of getting a sample containing a proportion defective greater than the action limit. Consequently, when such a sample is obtained, it is taken as an indication of an increase in the population proportion defective. The process is stopped, and the cause of the deterioration found and remedied.

Notice that, in the case of p-charts, lower limits are only of interest in so far as any significant decrease in proportion defective is very welcome, and steps should be taken to consolidate the improvement. It is usual to plot only upper limits. Notice too that, when controlling by dimensions, we had two charts ($\bar{x}$ and w) for each dimension, but we now have only one chart (p-chart) for each attribute.

Control charts for proportion defective may be based either on the proportion defective over the output of a complete day (or shift) or on the proportion defective found in samples from the output taken at intervals. In the former case the effective sample size (i.e., day or shift total output) may well vary, while in the latter case it should be constant. We must now consider the effect on p-charts of these two minor differences in circumstances.

a. ***p*-Charts. Constant Sample Size** If the sample size is constant, or nearly so, then the value of n in the expression

$$\bar{p}+3{\cdot}09\sqrt{\frac{\bar{p}\bar{q}}{n}}$$

may be taken to be the average size ($\bar{n}$) of the samples used to set up the charts. Thus:

$$\text{Action limit for } p = \bar{p}+3{\cdot}09\sqrt{\frac{\bar{p}\bar{q}}{\bar{n}}} \tag{9.1}$$

$$\text{Warning limit for } p = \bar{p}+1{\cdot}96\sqrt{\frac{\bar{p}\bar{q}}{\bar{n}}} \tag{9.2}$$

For example:

Day No.	1	2	3	4	5	6	7	8	9	10
No. of castings produced	154	152	148	150	154	145	151	154	150	153
No. of castings found to be defective	4	2`	2	4	3	4	2	2	1	4

Using each day's total output as the day's 'sample' draw up a control chart for proportion defective, on the basis of the proportion of defective castings produced in the ten days tabulated above.

Total number of defectives = 28
Total number of castings = 1511

$$\bar{p} = \frac{28}{1511} = 0{\cdot}0185 \qquad \bar{q} = 1-\bar{p} = 0{\cdot}9815$$

$$\bar{n} = \frac{1511}{10} = 151{\cdot}1$$

$$\text{Standard deviation of } p = \sqrt{\frac{\bar{p}\bar{q}}{\bar{n}}} = \sqrt{\frac{0{\cdot}0185\times 0{\cdot}9815}{151{\cdot}1}}$$

$$= 0{\cdot}00109$$

$$2\sqrt{\frac{\bar{p}\bar{q}}{\bar{n}}} = 0{\cdot}0022$$

$$\text{and } 3\sqrt{\frac{\bar{p}\bar{q}}{\bar{n}}} = 0{\cdot}0033$$

Hence

$$\tfrac{1}{40}\text{ (Warning) limit for } p = \bar{p}+2\sqrt{\frac{\bar{p}\bar{q}}{\bar{n}}}$$

$$= 0{\cdot}0185+0{\cdot}0022 = 0{\cdot}0207$$

$$\tfrac{1}{1000}\text{ (Action) limit for } p = \bar{p}+3\sqrt{\frac{\bar{p}\bar{q}}{\bar{n}}}$$

$$= 0{\cdot}0185+0{\cdot}0033 = 0{\cdot}0218$$

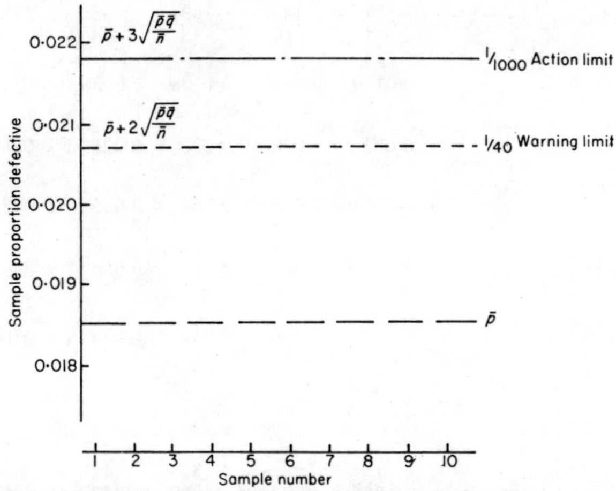

Fig. 9.1 Control chart for proportion defective—p-chart (sample size, n, approximately constant)

b. ***p*-Charts. Varying Sample Size** If the sample size (n) varies considerably, then the value of the standard deviation of p, i.e., $\sqrt{(\bar{p}\bar{q}/n)}$ will vary too, increasing as n decreases and vice-versa. Consequently the Control limit values will depend on the sample size and the Control Chart will no longer consist of a set of parallel lines.

To cater for this variation of Control limit we first plot a graph of

$$\text{Action Limit} = \bar{p}+3\sqrt{\frac{\bar{p}\bar{q}}{n}} \quad (9.3)$$

and

$$\text{Warning Limit} = \bar{p}+2\sqrt{\frac{\bar{p}\bar{q}}{n}} \quad (9.4)$$

against sample size n.

This graph is then used to read off the limits appropriate to any given sample size (n) as required. The limit lines on the Control Chart will then move up and down as the sample sizes decrease and increase. For example:

Batch No.	1	2	3	4	5	6	7	8	9	10
No. of pieces in batch	151	208	174	212	181	162	199	205	220	166
No. of defectives in batch	8	7	4	5	10	5	7	6	8	4

Construct a p-chart based on these data, and plot the ten batches on it.

Total number of pieces = 1878

Total number of defectives = 64

Therefore $\bar{p} = \dfrac{64}{1878} = 0{\cdot}034$

and $\bar{q} = 0{\cdot}966$

$\sqrt{(\bar{p}\bar{q})} = 0{\cdot}181$

Sample Size (n)	121	144	169	196	225	256
$\sqrt{n}$	11	12	13	14	15	16
$\sqrt{\frac{\bar{p}\bar{q}}{n}}$	0·0165	0·0151	0·0139	0·0129	0·0121	0·0113
$2\sqrt{\frac{\bar{p}\bar{q}}{n}}$	0·033	0·030	0·028	0·026	0·024	0·023
$3\sqrt{\frac{\bar{p}\bar{q}}{n}}$	0·050	0·045	0·042	0·039	0·036	0·034
Warning: $\bar{p}+2\sqrt{\frac{\bar{p}\bar{q}}{n}}$	0·067	0·064	0·062	0·060	0·058	0·057
Action: $\bar{p}+3\sqrt{\frac{\bar{p}\bar{q}}{n}}$	0·084	0·079	0·066	0·073	0·070	0·068

Batch No.	1	2	3	4	5
Propn. defective p	0·053	0·034	0·023	0·023	0·055
Batch No.	6	7	8	9	10
Propn. defective p	0·031	0·035	0·029	0·036	0·024

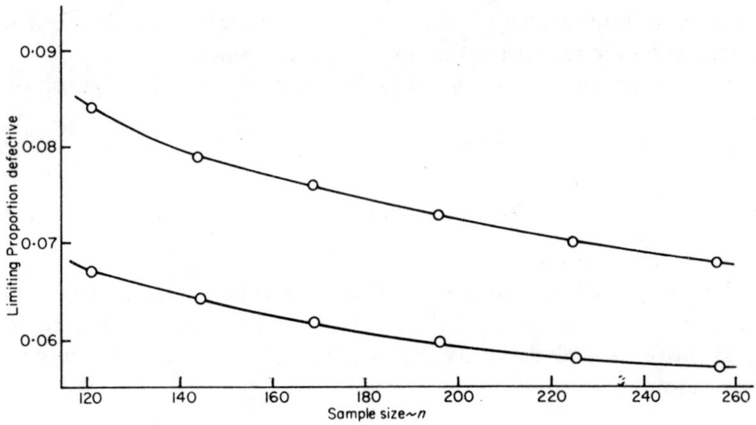

Fig. 9.2 Graph to determine warning and action limits for sample proportion defective (sample size n varying)

N.B. If sample sizes are approximately constant and $\bar{n}$ is used instead of varying n, then any point falling close to the limit line should be compared with the *true* limit for that particular sample size, i.e., $\bar{p}+3\sqrt{(\bar{p}\bar{q}/n)}$ instead of $\bar{p}+3\sqrt{(\bar{p}\bar{q}/\bar{n})}$. This may have the effect of bringing inside the 'correct' limit a point which was outside the approximate limit, or vice-versa.

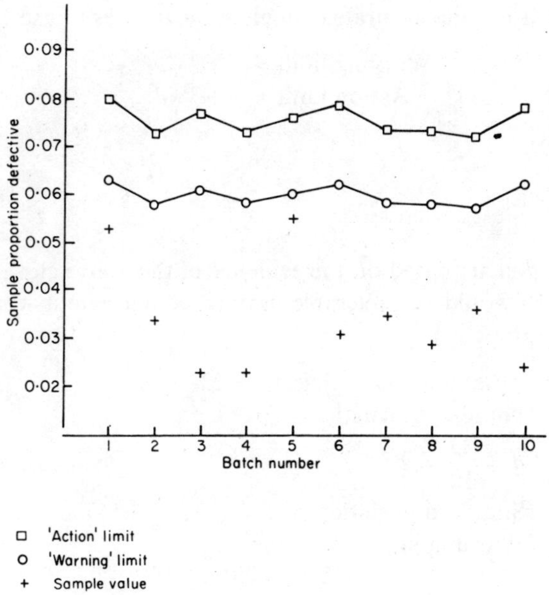

Fig. 9.3 Control chart for proportion defective—p-chart (sample size, n, varying)

9.3 *d*-charts

Since the probability of a defect occurring at any particular location in a certain piece is small, but the number of locations where a defect could occur is vast (e.g., a minor defect which could occur almost anywhere in a 100 metre length of cloth), the number of defects per unit (d) will form a Poisson distribution. The properties of the Poisson distribution, as shown in eq. (6.9) are such that the mean and the variance are numerically equal, hence if

$$\text{Mean number of defects per unit} = \bar{d}$$

then

$$\text{Standard deviation of number of defects per unit} = \sqrt{\bar{d}}.$$

Now, if $\bar{d} > 5$ the Normal is a good approximation to the Poisson (see section 6.6) and, even when $\bar{d} < 5$ the Normal is a fair approximation to the upper tail of the Poisson, hence we can use as our $\frac{1}{40}$ and $\frac{1}{1000}$ Control limits for d:

$$\text{Warning limit} = \bar{d} + 1{\cdot}96\sqrt{\bar{d}} \qquad (9.5)$$

$$\text{Action limit} = \bar{d} + 3{\cdot}09\sqrt{\bar{d}} \qquad (9.6)$$

As before, it is quite accurate enough in most cases to use

$$\text{Warning limit} = \bar{d}+2\sqrt{\bar{d}}$$
$$\text{Action limit} = \bar{d}+3\sqrt{\bar{d}}$$

For example:

100 m. cloth length No.	1	2	3	4	5	6	7	8	9	10
No. of minor defects found on inspection	2	3	1	4	4	0	2	1	4	2

Prepare a *d*-chart based on the evidence of the above cloth lengths. (More than 10 would be preferable in practice but here it would only add to the arithmetic.)

Total number of defects = 23
Number of lengths = 10

$$\bar{d} = \frac{23}{10} = 2{\cdot}3$$

$$\text{Standard deviation of } d = \sqrt{2{\cdot}3} = 1{\cdot}517$$
$$\text{Warning limit} = \bar{d}+2\sqrt{\bar{d}}$$
$$= 2{\cdot}3+3{\cdot}03 = 5{\cdot}3$$
$$\text{Action limit} = \bar{d}+3\sqrt{\bar{d}}$$
$$= 2{\cdot}3+4{\cdot}6 = 6{\cdot}9$$

If 7 or more defects are found in a 100 m length of cloth the quality of the material has probably deteriorated from its original state. Notice again that, as for *p*-charts, we only have upper limits on *d*-charts, since a decrease in *d* is an improvement.

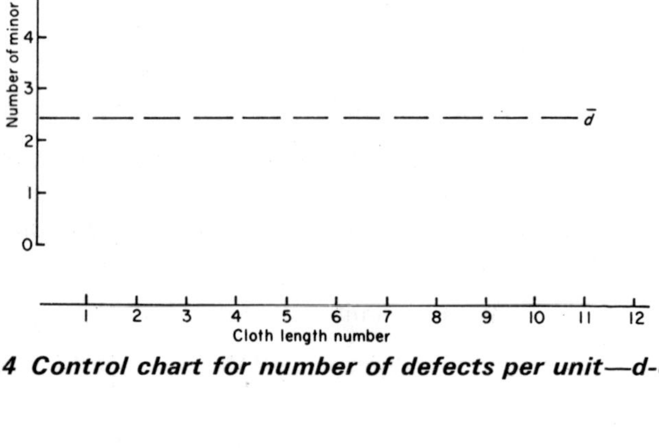

Fig. 9.4 Control chart for number of defects per unit—d-chart

9.4 Interchangeability of *p*- and *d*-charts

The sample size (n) on which a p-chart is based must, obviously, be reasonably large; again, one would hope that the proportion of defectives in the output (p) would be small, hence the Poisson distribution provides a good approximation to the theoretically more correct Binomial probabilities of obtaining samples containing any given proportion of defectives. Consequently, provided that the sample size is constant, a d-chart may be plotted instead of a p-chart by regarding d as 'number of defectives per sample' instead of 'number of defects per unit'. The reverse, plotting a p-chart instead of a d-chart cannot be done since 'number of defects per unit' cannot be expressed as a proportion, the total possible being theoretically infinite.

9.5 Problems

1. Make a control chart for the fraction defective (p chart) from the following data. Calculate the $\frac{1}{1000}$ control limit and draw the chart, showing the given points plotted. (N.B. the samples are roughly the same size so use $\bar{n}$.)

Week No.	No. of Magnets Inspected	No. of Defective Magnets	Fraction Defective p
1	724	48	0·0663
2	763	83	0·1088
3	748	70	0·0936
4	748	85	0·1136
5	724	45	0·0622
6	727	56	0·0770
7	726	48	0·0661
8	719	67	0·0932
9	759	37	0·0487
10	745	52	0·0698
11	736	47	0·0639
12	739	50	0·0677
13	723	47	0·0650
14	748	57	0·0762
15	770	51	0·0662
16	756	71	0·0939
17	719	53	0·0737
18	757	34	0·0449
19	760	29	0·0382
Totals	14,091	1,030	

2. Make a control chart for percentage defective from the following (N.B. varying n). Use the first 10 samples for calculating the limits and plot all 12 points on the chart.

Day	1	2	3	4	5	6	7
Output	162	182	174	180	125	182	143
Number Defective	6	11	6	4	4	5	2

Day	8	9	10	11	12
Output	172	175	163	169	193
Number Defective	5	5	4	3	6

3. Make a control chart for numbers of defectives (d chart) from the following data, using the first 20 samples for calculating the limits. All the samples are of the same size. Plot all 25 points on the chart.

Number of defectives per sample 9 10 7 11 12
8 11 17 10 13 8 10 8 11 3 10 6
15 9 11 12 11 4 11 8.

4. Samples of 40 articles are selected at regular intervals from the output of a stamping machine. The following list shows the number of defective articles in each of twenty such samples.

1, 0, 4, 0, 0, 4, 4, 1, 4, 2, 3, 3,
2, 3, 0, 2, 4, 2, 0, 1

Use these data to set up a suitable control chart for controlling the quality of future output.

5. The following data refer to the results of a visual inspection of castings. Each lot represents one day's production. The table shows the number of unacceptable castings produced in 30 consecutive lots.

Lot No.	1	2	3	4	5	6
No. inspected	135	162	140	155	188	166
No. of defectives	11	19	9	14	2	16

Lot No.	7	8	9	10	11	12
No. inspected	138	144	161	158	142	150
No. of defectives	10	12	11	14	8	9

Lot No.	13	14	15	16	17	18
No. inspected	163	136	160	126	147	159
No. of defectives	13	9	14	19	12	15

Lot No.	19	20	21	22	23	24
No. inspected	133	146	152	148	145	156
No. of defectives	11	16	8	18	12	9

Lot No.	25	26	27	28	29	30
No. inspected	153	144	140	173	135	145
No. of defectives	9	20	13	26	8	12

Use the first 10 lots to set up a control chart for fraction defective, putting in average control limits. Plot the next 20 points on the chart and comment on the condition of control, using more accurate control limits for such individual points as you consider necessary.

10. Confidence limits and significance tests

10.1 Introduction

Just because a random sample of 5 ball bearings from a very large batch, all of the same nominal size, has a mean diameter of 0·4997 in., we cannot say with certainty that the mean diameter of the whole batch would also be 0·4997 in. We would do better to say that the mean diameter of the batch lies between, perhaps, 0·4994 and 0·5000 in. Even this would not really be the whole story; we could similarly claim that the batch mean diameter lies between 0·4995 and 0·4999 in., or any other tolerance either side of the sample mean. Obviously we would have a better chance of being correct when we quoted the wider range of values, that is to say we would have more confidence in the estimate 0·4994 to 0·5000 in. than in the estimate 0·4995 to 0·4999 in. When we say that 95 per cent confidence limits for batch mean diameter are 0·4995 to 0·4999 in., and that 99·7 per cent confidence limits are 0·4994 to 0·5000 in., we mean that there is only a 5 per cent chance of obtaining a *sample* mean of 0·4997 in. (i.e., the one found) if the *batch* mean does not fall within the band 0·4995 to 0·4999 in.; similarly there is only a 0·3 per cent chance if the batch mean is not, in fact, inside the range 0·4994 to 0·5000 in.

The width of the interval, for a given confidence level, will depend on the variability (standard deviation) of the dimension, the greater the variability the wider the limits must be for any given level of confidence. It will also depend on the sample size, the greater the number of pieces the narrower the limits may be set for a given level of confidence.

Again, merely because two small samples, one from each of two—nominally the same—batches of ball bearings, yield sample mean diameters of 0·4998 and 0·5001 in. respectively, we cannot conclude that there would be a difference of 0·0003 in. between the mean diameters of the two batches. Indeed we cannot conclude that there is any difference at all between the batch mean diameters. Our task in this chapter will be to determine how great the difference between the sample means must be for it to be significant, i.e., to suggest that it is unreasonable to suppose that there is no difference between the *batch* means. The magnitude of the 'significant' difference will depend on three factors, the size of the samples—the larger the samples the smaller the difference needs to be to be significant; the variability (standard deviation) of the dimension—the greater the variability the greater the difference must be to be significant; and the significance level—if we say that the difference is significant at the 5 per cent level we mean that there is only a 5 per cent or less chance of obtaining two samples whose means differ by as much as, or more than, the means of the two random samples taken when sampling from identical batches.

There are three variations to be considered within the basic theme of significance testing. The first is the testing of a sample mean against an assumed population mean, e.g., if a sample mean diameter is 0·4997 in., is this sufficient evidence to suggest that the batch mean is *not* 0·5000 in. Secondly, the comparison of two unpaired samples, similar to those considered in the preceding paragraph. Lastly, the comparison of two paired samples, e.g., ten specimen lengths of mild-steel bar were each cut in half, one half of each specimen was subjected to the usual heat treatment and the other half to a treatment under development. The tensile strengths of the twenty pieces were measured, and noted, after heat treatment. Does the difference between the strengths of the two halves of each specimen suggest that it is unreasonable to say that the new treatment is no better than the old?

Methods of determining confidence intervals and performing significance tests will be explained by means of worked examples during the course of this chapter. First, however, we must state a number of theoretical relationships.

10.2 Theory required in significance testing and to compute confidence limits

a. **Unbiased Estimates of Population Parameters Based on Sample Observations** An 'unbiased' estimate is one which is as likely to be too great as it is to be too small. If many random samples (each of the same number of pieces) are taken from an infinite population and the mean of each sample is determined, then the mean of the sample means will be equal to the population mean, i.e., sample mean ($\bar{x}$) is an unbiased estimate of population mean (μ).

The mean of the standard deviations (s) of many random samples (all of the same number of pieces) will *not* be equal to the population standard deviation (σ). However, if we multiply s by $\sqrt{[n/(n-1)]}$, where n is the number of pieces in the sample, and call this $\hat{s}$, that is,

$$\hat{s} = s\sqrt{\frac{n}{n-1}} \tag{10·1}$$

then the mean value of $\hat{s}$ for all the samples *will* be equal to σ, i.e., *$\hat{s}$ is an unbiased estimate of population standard deviation σ, where $\hat{s} = s\sqrt{[n/(n-1)]}$ and s is the sample standard deviation.*

b. **Standard Deviation of the Distribution of the Sum or Difference of two Variates** If x and y are variates which are independently Normally distributed about means μ_x and μ_y with standard deviations σ_x and σ_y, and z is a variate formed by adding together a random value of x and a random value of y, that is

$$z = x+y$$

then z is a Normal variate and the mean value of z (μ_z) is the sum of the mean values of x and of y:

$$\mu_z = \mu_x+\mu_y \tag{10·2}$$

but the variance of z (σ_z^2) is the sum of the variances of x and of y:

$$\sigma_z^2 = \sigma_x^2+\sigma_y^2 \tag{10·3}$$

(N.B. It is *not* true to say that the standard deviation of z is equal to the sum of the standard deviations of x and of y.)

Again, if w is a variate formed by subtracting a random value of y from a random value of x:

$$w = x-y$$

then w is also a Normal variate and the mean value of w (μ_w) is the difference between the mean values of x and of y:

$$\mu_w = \mu_x-\mu_y \tag{10·4}$$

and the variance of w is the sum of the variances of x and of y:

$$\sigma_w^2 = \sigma_x^2+\sigma_y^2 \tag{10·5}$$

(N.B. Particular care must be taken to remember that, although w is formed by the *subtraction* of a value of the variate y from a value of the variate x, the variance of w is the *sum* of the variances of x and y.)

Example. A certain type of laminated board is produced by bonding 2 sheets of material A between 3 sheets of material B. The sheets are mass produced and it is found that A has a mean thickness of 0·25 mm with a standard deviation of 0·015 mm, B has a mean thickness of 0·15 mm with a standard deviation of 0·01 mm. The adhesive (of which 4 layers are used) has a mean thickness of 0·06 mm and standard deviation of 0·002 mm.

The boards are made up of sheets chosen at random from stock. Determine the mean and standard deviation of the thickness of the boards produced in this way.

Let suffix 'c' denote the adhesive. Then $\bar{x}_a$, $\bar{x}_b$, $\bar{x}_c$, $\bar{x}_F$ denote the mean thicknesses of materials, A, B, adhesive C, and finished board respectively. Similarly σ_a, σ_b, σ_c, σ_F denote the corresponding standard deviations.

Now

$$\bar{x}_F = 2\bar{x}_a + 3\bar{x}_b + 4\bar{x}_c \qquad \text{(eq. (10.2))}$$

that is,

$$\begin{aligned}
\bar{x}_F &= (2 \times 0{\cdot}25) + (3 \times 0{\cdot}15) + (4 \times 0{\cdot}06) \\
&= 0{\cdot}50 + 0{\cdot}45 + 0{\cdot}24 \\
&= 1{\cdot}19 \text{ mm.} \\
\sigma_F^2 &= 2\sigma_a^2 + 3\sigma_b^2 + 4\sigma_c^2 \qquad \text{(eq. (10.3))} \\
&= 2(0{\cdot}015)^2 + 3(0{\cdot}010)^2 + 4(0{\cdot}002)^2 \\
&= [(2 \times 225) + (3 \times 100) + (4 \times 4)] \times (0{\cdot}001)^2 \\
&= (450 + 300 + 16) \times (0{\cdot}001)^2
\end{aligned}$$

Therefore

$$\begin{aligned}
\sigma_F &= 0{\cdot}001\sqrt{766} \\
&= 0{\cdot}001 \times 27{\cdot}7 \\
&= 0{\cdot}028 \text{ mm approx.}
\end{aligned}$$

Mean thickness of laminated boards = 1·19 mm
and standard deviation = 0·028 mm.

c. **Estimate of Population Variance Based on Two Samples** If two samples, of n_1 and n_2 pieces respectively, are drawn from the same population (whose mean is μ and standard deviation is σ) and yield sample standard deviations s_1 and s_2 respectively, then the unbiased estimate ($\hat{s}$) of the population standard deviation (σ) is given by

$$\hat{s} = \sqrt{\frac{n_1 s_1^2 + n_2 s_2^2}{(n_1 - 1) + (n_2 - 1)}} \qquad (10.6)$$

d. **Distribution of Sample Means** If many samples, each of n pieces, are taken from a population whose mean is μ and standard deviation σ, and the mean of each sample calculated, then the sample means will themselves form a distribution, known as 'the distribution of sample means'. This new distribution will have a standard deviation, known as 'the standard error of the mean' which is given by:

$$\text{Standard Error (S.E.) of Sample Mean} = \sigma/\sqrt{n} \qquad (10{\cdot}7)$$

It is a property of the sample means that they are, approximately, Normally distributed whatever the shape of the distribution from which the samples were taken. The approximation improves as the sample size (n) becomes larger.

e. **'Student's t' distribution** We saw, in section 6.4, that if x_i is a member of a Normal population whose mean is μ and standard deviation is σ, and we write

$$z_i = \frac{x_i - \mu}{\sigma}$$

then z_i is also a Normal variate.

If we do not know the value of σ and have to use the 'estimated standard error' of x instead, then the new statistic

$$\frac{x_i - \mu}{\text{Estimated S.E. of } x}$$

is known as 'Student's t' and has a probability distribution of its own:

$$t = \frac{x - \mu}{\text{Estimated S.E. of } x} \qquad (10.8)$$

Every sample taken from a population is liable to produce a different estimate of the S.E. of the variate, consequently the denominator in the expression for t is a variable, while the population standard deviation σ is a constant so the denominator in the expression for z is a constant. Consequently, although the distribution of z follows a unique curve, the distribution of t varies according to the size of sample used to estimate the standard error. To be more precise, there is a different distribution curve for t for each number of 'degrees of freedom'; it will be sufficient to say—for our purposes—that if we use eq. (10.1) to estimate the standard error then t will have $n-1$ degrees of freedom, but if we have to use eq. (10.6) then t will have $(n_1-1)+(n_2-1)$ degrees of freedom.

The bigger the sample used to obtain the estimate of standard error the more 'reliable' is the estimate obtained, that is to say the variability of the denominator in eq. (10.8) decreases. The consequence of this is that, if the number of degrees of freedom is large, the t distribution and the standardized Normal distribution are the same.

Table D gives the values of t which have a probability Q of being exceeded by random sampling, for various degrees of freedom (v), for example, $v = 6$, $Q = 0{\cdot}05$; from Table D we find $t = 1{\cdot}943$. This means that, for a t distribution having 6 degrees of freedom, there are only 5 per cent of the population whose values are 1·943 or more. Similarly, another 5 per cent will have values $-1{\cdot}943$ or less. Consequently 10 per cent ($2Q$ shown at head of Table D) will fall outside $t = \pm 1{\cdot}943$. The former case is known as a 'one-tailed' t test and the appropriate probability level is indicated by Q. The latter is a 'two-tailed' test and the probability level is indicated by $2Q$.

f. **Variance Ratio (F) Test** Before we combine or compare the standard errors of two samples we must first satisfy ourselves that their respective variances are not so different that it is very unlikely that the two samples were drawn from populations having identical properties.

To do this we use the variance ratio (F) distribution.

If two samples from the same Normal population give unbiased estimates $\hat{s}_1$ and $\hat{s}_2$ respectively of σ, the population standard deviation, and the sample sizes are n_1 and n_2, then the ratio of the sample estimates of variance is known as F:

$$F = \frac{\hat{s}_1^2}{\hat{s}_2^2}$$

The larger estimate is always used as numerator, i.e., $\hat{s}_1 \geqslant \hat{s}_2$, so that F is never less than unity.

F has $v_1 = n_1 - 1$ and $v_2 = n_2 - 1$ degrees of freedom.

Table E shows the values of F which have given probabilities P of being exceeded in comparing random samples taken from the *same* population. The value of F corresponding to any given probability level P depends on the combination of degrees of freedom of numerator and denominator, v_1 and v_2.

g. **Conventional Significance Levels** When significance testing it is first necessary to set up a 'hypothesis' to be tested. The hypothesis may be 'the two machines from which the two samples were taken are not set to different dimensions' or 'the setting of the two machines differs by 0·020 in.' Having set up the hypothesis we then assume it to be true and determine the probability of obtaining, by chance, the actual figures obtained in the samples.

If this probability is 5 per cent or less we say that the difference is significant, i.e., the hypothesis is unlikely to be true—so we reject it.

1 per cent or less is *Highly Significant* and
0·1 per cent or less is *Very Highly Significant*

Care must be taken here to realize that we cannot say that the hypothesis *is false*, we can only say that it is *not likely* to be true, in the light of the available evidence. More evidence, for example larger samples, could cause us to alter our decision.

10.3 Worked examples

a. **Example 1. Confidence Limits—Population Standard Deviation Known** Over a long period of time it has been well established that the standard deviation of length of a component, mass-produced on a particular machine, is 0·002 in. After resetting, a sample of 100 components yielded a mean length of 0·4607 in.

Determine
(a) 95 per cent and
(b) 99·7 per cent,
confidence limits for the mean length of the components produced at this setting.

Solution: Let the unknown population mean be μ. Let the sample mean be $\bar{x}$.

Now the distribution of sample means is approximately Normal, having a mean value equal to the population mean (i.e., mean value of $\bar{x}$ is μ) and the standard deviation of the mean is $\sigma/\sqrt{n}$, where n is the sample size (i.e., number of pieces in the sample) and σ is the population standard deviation.

Hence, if we write

$$z = \frac{\bar{x}-\mu}{\sigma/\sqrt{n}}$$

z is a Normal variate distributed about mean μ with standard deviation $\sigma/\sqrt{n}$.

From Table C we see that there is only a 5 per cent chance of z being greater than 1·96 or less than $-1{\cdot}96$; and only a 0·3 per cent chance that z will be greater than 3 or less than -3. But $\bar{x}$ (sample mean) and $\sigma/\sqrt{n}$ (standard deviation of sample mean) are constants. Consequently there is only a 5 per cent chance that μ will fall outside the range $\bar{x} \pm 1{\cdot}96\,\sigma/\sqrt{n}$; and only a 0·3 per cent chance that it will fall outside $\bar{x} \pm 3\sigma/\sqrt{n}$.

Consequently we say that 95 per cent confidence limits for population mean μ are

$$\bar{x} \pm 1{\cdot}96\sigma/\sqrt{n}$$

that is $0{\cdot}4607 \pm 1{\cdot}96 \times \dfrac{0{\cdot}002}{10}$

which is $0{\cdot}4607 \pm 0{\cdot}0004$ in.
or 95 per cent confidence limits for component length are $0{\cdot}4607 \pm 0{\cdot}0004$ in.

Similarly, 99·7 per cent confidence limits for components length are $0{\cdot}4607 \pm 3 \times 2 \times 10^{-4}$, i.e., $0{\cdot}4607 \pm 0{\cdot}0006$ in.

This indicates that, if every component produced—by this machine at this setting—was to be measured and the population mean length evaluated, there is only a 5 per cent chance that it would differ from 0·4657 in. by more than 0·0004 in., and only 0·3 per cent chance that the difference would be more than 0·0006 in.

If the same 'mean' had been obtained from a smaller sample (say only 4 pieces) we could not have put so much confidence in our estimate. In this case:

$$\text{Standard deviation of sample mean} = \sigma/\sqrt{n}$$

$$= \frac{0{\cdot}002}{\sqrt{4}} = \frac{0{\cdot}002}{2} = 0{\cdot}001 \text{ in.}$$

Therefore 95 per cent confidence limits for mean length would be $0{\cdot}4607 \pm 1{\cdot}96 \times 0{\cdot}001$, that is, $0{\cdot}4607 \pm 0{\cdot}002$ in. approximately (compared with $0{\cdot}4607 \pm 0{\cdot}0004$ in. based on sample of 100).

Similarly, 99·7 per cent confidence limits would be $0{\cdot}4607 \pm 0{\cdot}003$ in. compared with $0{\cdot}4607 \pm 0{\cdot}0006$ based on sample of 100 pieces.

b. **Example 2. Confidence Limits—Population Standard Deviation Unknown** We consider again the previous problem, but this time let us assume that the population standard deviation is *not* well established but that the standard deviation of the length of the 100 pieces measured is 0·002 in.

Population standard deviation σ is unknown but sample standard deviation $s = 0{\cdot}002$ in.

In this case we can no longer consider

$$z = \frac{\bar{x}-\mu}{\sigma/\sqrt{n}}$$

since σ is unknown and we must use

$$t = \frac{\bar{x}-\mu}{\hat{s}/\sqrt{n}}$$

Now $\hat{s} = s\sqrt{\dfrac{n}{n-1}}$

But $\quad n = 100$, therefore

$$\sqrt{\frac{n}{n-1}} = 1 \text{ approximately}$$

and $\quad \hat{s} = s$ approximately

and $\quad \hat{s}/\sqrt{n} = s/10$

$$= 0{\cdot}002/10 = 0{\cdot}0002 \text{ in.}$$

and the distribution of t is similar to the Normal distribution when the sample, and hence the number of degrees of freedom, is large (in this case $v = n-1 = 99$). Consequently the rest of the solution to this problem is identical to that in example 1.

However, had the sample consisted of 4 instead of 100 pieces, we would have had:

$$\begin{aligned}\hat{s} &= s\sqrt{\frac{n}{n-1}} \\ &= 0{\cdot}002\times\sqrt{\tfrac{4}{3}} \\ &= 0{\cdot}002\times 1{\cdot}153 = 0{\cdot}0023\end{aligned}$$

Now $$t = \frac{\bar{x}-\mu}{\hat{s}/\sqrt{n}}$$

therefore $$\begin{aligned}\bar{x}-\mu &= t\left(\frac{\hat{s}}{\sqrt{n}}\right) \\ &= t\times\frac{0{\cdot}0023}{2} \\ &= 0{\cdot}00115\,t\end{aligned}$$

From Table D, for a t distribution having $v = 4-1 = 3$ degrees of freedom, $2Q = 0{\cdot}05$, we see that $t = 2{\cdot}353$; i.e., only 5 per cent of the values making up the t distribution having $v = 3$ will fall outside $t = \pm 2{\cdot}353$.

Thus there is only a 5 per cent chance that μ will fall outside

$$\bar{x}\pm 0{\cdot}00115\times 2{\cdot}353$$

That is, 95 per cent confidence limits for population mean are 0·4607 ±0·0027 in.

Notice that the interval is wider (0·0027 instead of 0·002 in.) than when the population standard deviation was known. This is, of course, what we would expect as a result of the added uncertainty due to estimating population standard deviation.

c. **Significance Test for Difference between Sample Mean and Hypothetical Population Mean** Let us now suppose that the samples considered in our earlier examples were taken from the output of a machine producing to a nominal dimension of 0·4600 in. Our problem now is to decide if the sample mean, which was found to be 0·4607 in., could have arisen by pure chance, even though the population mean is 0·4600, or if 0·4607 in. indicates that the machine is probably set slightly oversize.

Referring back to the previous calculations, and tabulating, we have the following information:

From Example 1:

No. of pieces in sample	Confidence Limits (in.) 95 per cent	99·7 per cent
100	0·4607 ±0·0004	0·4607 ±0·0006
4	0·4607 ±0·002	0·4607 ±0·003

From Example 2:

No. of pieces in sample	Confidence Limits (in.) 95 per cent	99·7 per cent
100	0·4607 ±0·0004	0·4607 ±0·0006
4	0·4607 ±0·0027	

This shows that if the 0·4607 in. sample mean was obtained from a sample of 100 pieces, then the hypothetical population mean of 0·4600 in. is outside both 95 per cent and 99·7 per cent confidence limits.

Consequently we say that the difference of 0·0007 in. from nominal is significant at the 5 per cent and 1 per cent levels (in fact 0·3 per cent level) and reject the hypothesis, i.e., the sample evidence suggests that the population mean is not 0·4600 in.

However, if the sample mean of 0·4607 in. was obtained from a sample of only 4 pieces, then the hypothetical mean of 0·4600 in. is well within the 95 per cent confidence limits and we say the difference from nominal is not significant at the 5 per cent level and accept the hypothesis, i.e., the sample evidence is not sufficient to cast doubt on the claim that the population mean is 0·4600 in.

To sum up, for a population whose standard deviation is 0·002 in. (either known or estimated) and mean is 0·4600 in., there is a less than 5 per cent chance of obtaining a sample of 100 pieces whose mean is 0·4607 in., but more than 5 per cent chance of obtaining a sample of 4 pieces having this mean. Hence if a sample of 100 pieces with this mean occurs it strongly suggests that the assumed population mean is incorrect, though a sample of 4 pieces with such a mean would not provide strong enough evidence to make the same suggestion.

d. **Significance Test for the Difference between the Means of Two Unpaired Samples** *Problem* Two machines are producing locating dowels to the same nominal length of 1·850 in. A sample of 9 dowels from machine A is found to have a mean length of 1·852 in. and a standard deviation of length of 0·003 in. A sample of 16 dowels from machine B is found to have a mean length of 1·849 in. and a standard deviation of length of 0·002 in.

Is the difference between the means sufficient to suggest that

(a) the machines are not set to produce the same dimension

(b) machine A is set to produce a longer dowel than machine B?

Solution

	Machine A	Machine B
Sample size	$n_A = 9$	$n_B = 16$
Sample mean	$\bar{x}_A = 1{\cdot}852$	$\bar{x}_B = 1{\cdot}849$
Sample standard deviation	$s_A = 0{\cdot}003$	$s_B = 0{\cdot}002$

$$\left.\begin{aligned}\hat{s}_A &= s_A\sqrt{\frac{n_A}{n_A-1}} = 0{\cdot}003\sqrt{\frac{9}{8}}\\ \hat{s}_B &= s_B\sqrt{\frac{n_B}{n_B-1}} = 0{\cdot}002\sqrt{\frac{16}{15}}\end{aligned}\right\} \qquad \text{(eq. (10.1))}$$

F ratio test: (page 120)

$$F = \hat{s}_A^2/\hat{s}_B^2 = \frac{9\times 10^{-6}}{4\times 10^{-6}}\times\frac{9}{8}\times\frac{15}{16}$$

$$= \frac{9}{4}\times\frac{9}{8}\times\frac{15}{16} = 2{\cdot}37$$

$$v_1 = n_A - 1 = 8 \quad v_2 = n_B - 1 = 15$$

From Table E, the value of F at the 5 per cent level, for these degrees of freedom, is 2·64. Consequently, our value of 2·37 is not significant, at the 5 per cent level; our sample variances are not sufficiently different to doubt the assumption that they are samples from identical populations.

Unbiased estimate of population standard deviation =

$$\hat{s} = \sqrt{\frac{n_A s_A^2 + n_B s_B^2}{(n_A-1)+(n_B-1)}} \qquad \text{(eq. (10.6))}$$

$$= \sqrt{\frac{(9\times 9\times 10^{-6})+(16\times 4\times 10^{-6})}{(9-1) \quad + \quad (16-1)}}$$

$$= \sqrt{\frac{81+64}{23}\times 10^{-6}} = 0{\cdot}0025$$

Estimated standard error of sample mean $= \dfrac{\hat{s}}{\sqrt{n}}$ (eq. (8.4))

Therefore estimated standard error of $\bar{x}_A$

$$= \frac{0{\cdot}0025}{\sqrt{9}} = \frac{0{\cdot}0025}{3}$$

and estimated standard error of $\bar{x}_B$

$$= \frac{0{\cdot}0025}{\sqrt{16}} = \frac{0{\cdot}0025}{4}$$

Now 'variance of a difference = sum of the variances' (eq. (10.5))
Therefore variance of difference between sample means

$$= \left(\frac{0{\cdot}0025}{3}\right)^2 + \left(\frac{0{\cdot}0025}{4}\right)^2$$

Estimated S.E. of difference between sample means

$$= \sqrt{\text{variance}}$$

$$= 0{\cdot}0025\sqrt{\left[\frac{1}{9}+\frac{1}{16}\right]} = 0{\cdot}0025\sqrt{\frac{16+9}{16\times 9}}$$

$$= 0{\cdot}0025\sqrt{\frac{25}{144}} = \frac{0{\cdot}0025\times 5}{12}$$

Now even if the individual dimension is not Normally distributed it is reasonable to assume an approximately Normal distribution for the sample means, hence also for the difference between sample means. Consequently

$$\frac{(\bar{x}_A-\bar{x}_B)-(\mu_A-\mu_B)}{\text{Estimated S.E. of }(\bar{x}_A-\bar{x}_B)} \text{ is a '}t\text{' variate.}$$

Hypothesis: Difference between population means $= \mu_A - \mu_B = 0$

Difference between sample means

$$= \bar{x}_A - \bar{x}_B$$
$$= 0{\cdot}003 \text{ in.}$$

Hence

$$t = \frac{0{\cdot}003-0}{0{\cdot}0025\times 5/12}$$

$$= \frac{3\times 12}{2{\cdot}5\times 5} = \frac{36}{12{\cdot}5} = 2{\cdot}88$$

and has $v = (n_A-1)+(n_B-1) = 23$ degrees of freedom.

To answer question (a) we require a 'two-tail' test since the question allows for a difference between the means in either direction. For question (b) we only regard a suggestion that machine B *could* be producing to a larger size than machine A as significant—the opposite being in agreement with our hypothesis (in this case that $\mu_A > \mu_B$)—and hence we use a one-tail test.

From Table D:

$v = 23$

(a) $2Q = 0{\cdot}05 \quad t = 2{\cdot}069$

(b) $Q = 0{\cdot}05 \quad t = 1{\cdot}714$

Our value of t ($= 2{\cdot}88$) is greater than each of these and is, therefore, significant in both cases at the 5 per cent level. The evidence of the samples is such as to suggest, at the 5 per cent level, that

(a) the machines are not set to produce the same dimension, and
(b) machine A is set to produce a longer dowel than machine B.

The words 'at the 5 per cent level' mean that there is less than 5 per cent chance of obtaining these samples if

(a) the machines are set to produce the same dimension; and less than 5 per cent chance of obtaining these samples if
(b) machine A is not set to produce a longer dowel than machine B.

e. **Significance Test for the Difference between Two Paired Samples**

Problem Ten specimen lengths of steel bar were each cut in half. One group of ten halves was subjected to heat treatment A and the other to heat treatment B. After heat treatment all 20 pieces were tested for tensile strength, and the results tabulated. Do these data suggest that:

(a) there is a difference in strength attributable to the nature of the heat treatment.
(b) heat treatment B produces a greater strength than treatment A?

Tensile strength lbf/in² $\times 10^6$

Specimen	1	2	3	4	5	6	7	8	9	10
Treatment A	20·2	18·8	22·1	19·7	19·5	20·1	21·1	20·0	22·1	18·4
Treatment B	20·3	20·1	21·7	20·4	21·5	20·8	20·9	22·1	21·9	20·3

Solution Here the samples are 'paired', that is to say for each specimen we know the strength after both treatments A and B. Consequently if we consider the distribution of the *differences* in strength produced by the two treatments we can eliminate the effect due to variations in basic material, before treatment.

We must assume that the strengths after each treatment are Normally distributed; then the distribution of differences will also be Normal.

Specimen	1	2	3	4	5	6
Difference in Strength (B – A)	0·1	1·3	–0·4	0·7	2·0	0·7

Specimen	7	8	9	10
Difference in Strength (B – A)	–0·2	2·1	–0·2	1·9

Let x denote the difference in strength. Then

$$\Sigma x = 8{\cdot}0 \quad n = 10$$

therefore $$\bar{x} = 0{\cdot}8$$

$$\Sigma x^2 = 15{\cdot}14$$

therefore Variance of $$x = \frac{15{\cdot}14}{10} - (0{\cdot}8)^2$$

$$= 1{\cdot}514 - 0{\cdot}64 = 0{\cdot}874$$

Standard deviation of $x = \sqrt{0{\cdot}874}$

Estimated population standard deviation

$$= \sqrt{0{\cdot}874} \times \sqrt{\frac{10}{9}} \quad = s\sqrt{\frac{n}{n-1}}$$

Estimated standard error of $\bar{x}$

$$= \sqrt{0{\cdot}874} \times \sqrt{\frac{10}{9}} \times \sqrt{\frac{1}{10}}$$

$$= \frac{1}{3}\sqrt{0{\cdot}874} = 0{\cdot}312$$

(a) *Hypothesis* Mean of population of strength differences (μ) = 0

$$t = \frac{\bar{x}-\mu}{\text{Estimated S.E. of } \bar{x}}$$

$$= \frac{0{\cdot}8-0}{0{\cdot}312} = 2{\cdot}56$$

Number of degrees of freedom = $v = 10-1 = 9$. From Table D; two-tail test $2Q = 0{\cdot}05$, $v = 9$, $t = 2{\cdot}262$.

Our value, $t = 2{\cdot}56$, is greater than the value of t at the 5 per cent level and is, consequently, significant. The sample data do not support the suggestion that there is no difference between the strengths produced by the two treatments.

(b) *Hypothesis* Mean of population of strength differences (μ) $\ngtr 0$.

To test this hypothesis we require a one-tailed test. From Table D, using $v = 9$ and $Q = 0{\cdot}05$, $t = 1{\cdot}833$.

Hence our value, $t = 2{\cdot}56$, is again significant and we reject the hypothesis. The sample data suggest that treatment B does produce a greater strength than treatment A.

10.4 **Problems**

1. A machine is supposed to produce steel rods of nominal length 3 in. and standard deviation 0·02 in. Do the data of the following sample give cause to doubt that the machine is keeping to specification? Rod lengths, in. 2·97, 3·02, 2·98, 2·98, 2·97.
2. Shafts are being produced automatically to a nominal diameter of 3·752 in. Nine shafts taken at random measured 3·756, 3·759, 3·754, 3·750, 3·755, 3·760, 3·753, 3·758, 3·749. Do these data give cause to doubt that the machine has been coreectly set?
3. Two machines are producing pieces to the same specification. A random sample of 10 pieces produced by machine A were checked on a particular dimension and the following differences from nominal found. (in. × 0·0001): 0, −4, +6, +7, +3, +2, −2, +4, +5, −1. A random sample of 12 pieces from machine B yielded the following differences from nominal (in. × 0·0001): −3, +3, +12, +5, +2, +4, +8, −2, +11, +13, 0, +7. Do these data give cause

to doubt the claim that there is no difference between the pieces produced by machine A and the pieces produced by machine B?

4. Following a work-study investigation a change in a certain operation was tried out. The times taken to carry out the operation, both by the old and by the new method, for each of 10 workers were measured. Do these observations support the claim that the new method is quicker?

Worker	A	B	C	D	E	F	G	H	J	K
Old time (sec)	27	28	25	34	30	29	29	31	38	32
New time (sec)	24	28	22	38	29	26	28	32	36	30

5. A sample of 100 electric light bulbs produced by manufacturer A showed a mean lifetime of 1190 hours and a standard deviation of 90 hours. A sample of 75 bulbs produced by manufacturer B showed a mean lifetime of 1230 hours and a standard deviation of 120 hours. Do these two samples support the claim of manufacturer B that his product has a longer mean life?
6. Measurements of the diameters of a random sample of 200 ball bearings made by a certain machine during one week showed a mean of 0·824 in. and a standard deviation of 0·042 in. Determine
   (a) 95 per cent,
   (b) 99 per cent,
   confidence limits for the mean diameter of all the ball bearings.
7. The known process standard deviation for a particular dimension, produced on a certain machine, is 0·004 in. How large a sample must be taken from the output of this machine if the sample mean is to be used as an estimate of the dimension set, and we wish to be
   (a) 95 per cent,
   (b) 99 per cent,
   confident that the error of our estimate will not exceed 0·0005 in.?
8. Two sources of raw material are under consideration. Both seem to have about the same average characteristics but there is some doubt about their variabilities being equal. A sample of 13 lots from A has a variance of 146 while a sample of 21 lots from B has a variance of 200. Is it likely that the two population variances are equal?

11. Chi-squared test

11.1 Introduction

Imagine a large quantity of resistors in a box. It is thought that 50 per cent are rated at 10 ohms, 20 per cent at 5 ohms, 20 per cent at 2 ohms, and 10 per cent at 1 ohm. A random sample of 50 resistors is taken and found to contain 28 rated at 10 ohms (we would have expected 25), 11 rated at 5 ohms (10 expected), 8 rated at 2 ohms (10 expected) and 3 rated at 1 ohm (5 expected). Our problem is to decide if the difference between the contents of the sample and what we would have 'expected' is great enough to give reason to doubt our initial assumption. To make this decision we employ the Chi-squared distribution. Chi is a Greek letter, pronounced ki, as in kite, and written χ. The Chi-squared distribution will be defined in the next paragraph. The values of χ^2 which have only a 5 per cent probability of being exceeded due to differences arising from sampling alone, i.e., if the assumption made initially is correct, are given in Table F.

11.2 χ^2 Distribution

If a trial can result in any one of n different 'events' and we use O_1, O_2, O_3, ... O_i ... O_n to denote the number of times Event 1, Event 2, Event 3, ... Event i ... Event n were observed to occur in a series of trials, and E_1, E_2, E_3 ... E_i ... E_n to denote the number of times we would 'expect' them to occur, assuming a given hypothesis to be true, then

$$\chi^2 = \frac{(O_1-E_1)^2}{E_1}+\frac{(O_2-E_2)^2}{E_2}+\frac{(O_3-E_3)^2}{E_3}+ \dots +\frac{(O_i-E_i)^2}{E_i}+\dots+\frac{(O_n-E_n)^2}{E_n}$$

and has $n-1$ degrees of freedom.

The reliability of this form of χ^2 increases as the number of trials increases (i.e., as O_i and E_i; $i = 1, 2, 3 \dots n$ increases) and should never be used when a value of E_i is less than 5. If it is necessary to test a set of observations for which one or more of the 'expected' frequencies falls below 5, the offending frequency must be combined with another group (or groups) to bring the expected frequency above 5. If this has to be done the number of degrees of freedom for χ^2 is reduced by one each time a small group is merged into another.

11.3 Example 1

Using the data set out in section 11.1 we have:

Resistance (ohms)	10	5	2	1	
Observed frequency (O)	28	11	8	3	Total = 50
Expected frequency (E)	25	10	10	5	Total = 50

$O-E$	3	1	−2	−2	Total* = 0
$(O-E)^2$	9	1	4	4	
$\frac{(O-E)^2}{E}$	0·36	0·1	0·4	0·8	Total = 1·66

* Since the sum of the 'observed' frequencies and the sum of the 'expected' frequencies are equal, $\sum(O-E)$ should be zero and provides a useful check.

Thus $\chi^2 = 1{\cdot}66$ and has $4-1 = 3$ degrees of freedom. From Table F, for 3 d.f., any value of χ^2 in excess of 7·81 is significant at the 5 per cent level, hence our value of 1·66 is not significant, i.e., there is a better than 5 per cent chance of obtaining such a sample as ours from a population containing 50 per cent at 10 ohms, 20 per cent at 5 ohms, 20 per cent at 2 ohms, and 10 per cent at 1 ohm, so we have no reason to doubt the claim that these are the proportions.

11.4 Example 2

Absence records for the AFY Company over a period of 30 working davs show the following figures.

Trade	No. of men employed	No. of man/days absence
Turner	24	7
Grinder	12	6
Miller	18	5
Fitter	34	10
Labourer	20	8
Total	108	36

Test the hypothesis that the absentee rate is independent of trade.

Average number of man/days lost per man, in the 30 day period is

$$\frac{36}{108} = \frac{1}{3}$$

If absenteeism is independent of trade we would expect the number of man/days lost by turners to be

$$24 \times \frac{1}{3} = 8$$

and by grinders to be

$$12 \times \frac{1}{3} = 4$$

etc.

Hence:

Trade	No. of men employed	No. of man/days absence Observed (O)	No. of man/days absence Expected (E)	$\frac{(O-E)^2}{E}$
Turner	24	7	8	0·125
Grinder	12	6	*4 }	0·100
Miller	18	5	6 }	
Fitter	34	10	$11\frac{1}{3}$	0·154
Labourer	20	8	$6\frac{2}{3}$	0·267
			Total	0·646

* Since the expected frequency is less than 5 this group must be combined with the one next to it.

$$\chi^2 = \Sigma\frac{(O-E)^2}{E}$$

$$= 0{\cdot}646$$

and has $4-1 = 3$ degrees of freedom (*not* $5-1 = 4$, since two trades were combined, leaving only 4 groups).

From Table F we see that the value of χ^2, with 3 degrees of freedom, which is significant at the 5 per cent level, is 7·81. Consequently the sample data give no cause to doubt the claim that absentee rate is independent of trade.

11.5 **Goodness of fit**

The χ^2 test can be used in a similar manner to test the 'goodness of fit' of an observed set of data to a given theoretical distribution. The 'goodness of fit' is tested by comparing the observed frequencies of each value of the variate with the frequencies given by the theoretical distribution. The parameters of the theoretical distribution (e.g., mean and standard deviation) are chosen to be the same as those of the observations. χ^2 loses one degree of freedom for each constraint placed on the theoretical frequencies (e.g., $v = n-3$) if the theoretical frequencies are so ordered that they have the same mean, standard deviation, and sum as the observed frequencies.

11.6 Reliability of sample estimate of population standard deviation

It can be shown that

$$\frac{(n-1)s^2}{\sigma^2},$$

where s is the standard deviation of a sample of n items drawn from a population whose (unknown) standard deviation is σ, is distributed like χ^2 with $(n-1)$ degree of freedom. Consequently we can use this to determine the greatest value σ is likely to have when the only information available is s, based on a sample.

a. **Example** We are required to make a quick decision as to the acceptability of a large batch of spacers. To be of any use to us the thickness must not vary by more than $\pm 0{\cdot}08$ mm.

A random sample of 10 spacers taken from the batch had the following thicknesses (difference from nominal in mm):

$+0{\cdot}02$	$0{\cdot}00$	$-0{\cdot}01$	$+0{\cdot}01$	$0{\cdot}00$
$-0{\cdot}02$	$+0{\cdot}01$	$+0{\cdot}03$	$0{\cdot}00$	$-0{\cdot}01$

First we calculate the sample variance (s^2)

Difference from nominal mm $\times 0{\cdot}01$ x	f	fx	fx^2
−2	1	−2	4
−1	2	−2	2
0	3	0	0
1	2	2	2
2	1	2	4
3	1	3	9
Total	10	7 − 4 = 3	21

$$\bar{x} = \frac{\Sigma fx}{\Sigma f} = \frac{3}{10} = 0{\cdot}3$$

$$s^2 = \frac{\Sigma fx^2}{\Sigma f} - (\bar{x})^2 = \frac{21}{10} - (0{\cdot}3)^2$$

$$= 2{\cdot}01$$

Sample variance $= 2{\cdot}01\ (\text{mm} \times 10^{-2})^2$

Now $[(n-1)s^2]/\sigma^2$ is distributed like χ^2 with $(n-1)$ degrees of freedom

$$v = n-1 = 9$$
$$P = 95 \text{ per cent}$$

From Table F $\quad \chi^2 = 3{\cdot}33$

That is, $9s^2/\sigma^2$ has a 95 per cent chance of exceeding 3·33 so we can be 95 per cent sure that σ^2 *will not* exceed $9s^2/3{\cdot}33$. We can be 95 per cent sure that population variance σ^2 will be less than

$$\frac{9 \times 2{\cdot}01}{3{\cdot}33} (= 5{\cdot}4)$$

95 per cent confidence limits for population tolerance

$$= \pm 3\sigma$$
$$= \pm 3\sqrt{5{\cdot}4}$$
$$= \pm 7 \text{ mm} \times 10^{-2}$$

so although we have only inspected 10 spacers we can be 95 per cent sure that the tolerance over the whole batch will not exceed $\pm 0{\cdot}07$ mm; since this is less than the stipulated $\pm 0{\cdot}08$ mm we will accept the batch.

b. **Example** Nominal 12-mm diameter rivets are being produced in large batches. Determine the size of sample we must take in order to be 95 per cent sure that the population standard deviation does not exceed (a) 2·0, (b) 1·2 times the sample standard deviation.

$$\frac{(n-1)s^2}{\sigma^2} = \chi^2$$

$$\sigma^2 = \frac{(n-1)s^2}{\chi^2}$$

$$= \frac{n-1}{\chi^2} s^2$$

therefore $\quad \sigma = s\sqrt{\frac{n-1}{\chi^2}}$

From Table F with $P = 0{\cdot}95$ we obtain the first two columns of the following table, and hence compute the next two columns.

$n-1 (=\nu)$	χ^2	$\frac{n-1}{\chi^2}$	$\sqrt{\frac{n-1}{\chi^2}}$
5	1·15	4·348	2·085
6	1·64	3·663	1·914
7	2·17	3·226	1·796
8	2·73	2·933	1·713
9	3·33	2·703	1·644
10	3·94	2·538	1·593
20	10·85	1·843	1·358
30	18·49	1·623	1·274
40	26·51	1·508	1·228
50	34·76	1·438	1·199
100	77·93	1·283	1·133

This table shows that $\sqrt{[(n-1)/\chi^2]}$ is first less than 2 when $n-1 = 6$, i.e., to be 95 per cent sure that σ is less than $2 \times s$ we need a sample of $n = 7$ pieces.

Again, $\sqrt{[(n-1)/\chi^2]}$ is first less than 1·2 when $n-1 = 50$, i.e., to be 95 per cent sure that σ is less than $1{\cdot}2 \times s$ we need a sample of $n = 51$ pieces.

If, therefore, we use a sample of 7 pieces we can be 95 per cent sure that the population standard deviation will not exceed twice the sample standard deviation, and for a sample of (say) 50 pieces we can be 95 per cent sure that σ does not exceed $1{\cdot}2s$.

11.7 Problems

1. A plant consists of 5 similar compressors. Over a period of years records show the following numbers of piston ring failures. Do these data give you sufficient evidence to doubt that the probability of a piston ring failure is the same for all the compressors?

Compressor	I	II	III	IV	V
No. of piston ring failures	48	32	36	50	34

2. Samples, each of 100 pieces, were taken from the ends of two production lines turning out the same product. The sample from line A was found to contain 15 defective and from line B 5 defective. Does this give cause to doubt that line A is as efficient as line B? Would your conclusions have been different if the numbers defective had been 18 and 2 respectively?
3. An illumination survey of a factory site showed the following figures for number of light fittings, using lamps of a given wattage, and the average number replaced per four-week period following failure.

Wattage	No. of fittings	Average monthly replacements
300	1200	90
200	1500	128
150	1000	90
100	820	64
15	2200	108

On the assumption that all the fittings have the same utilization, test the hypothesis that the rate of failure is independent of wattage.

4. Five moulds were used to produce similar, high-precision castings. Because of the great accuracy required a large number had to be rejected. The number produced from each mould, and the number rejected on inspection, are tabulated below. Is there any evidence to suggest that the moulds are not all equally good?

Mould	A	B	C	D	E	Total
No. made	10	9	7	13	11	50
No. rejected	8	8	3	7	9	35

5. Extrusions were produced using six different pressures but the same die. The number manufactured at each pressure, and the number rejected as faulty, are recorded below. Does the evidence cast reasonable doubt on the claim that extrusion quality is independent of pressure (over the range tested)?

Pressure lb in^{-2} $\times 10^3$	1·8	1·9	2·0	2·1	2·2	2·6	Total
No. manufactured	5	12	15	11	9	8	60
No. rejected	4	5	6	8	7	6	36

Revision problems

(Using S.I. units where applicable)

1. (a) A random sample of 20 pieces, taken from a large batch of mass-produced components, is inspected and found to contain 1 defective. Determine if it is more likely that the batch contains 3 per cent or 7 per cent of defectives.
   (b) If a second random sample, this time of 50 pieces, is taken from the same batch and found to contain 2 defectives, use all the information to decide which is the more likely batch proportion defective now.
2. Determine the probability that a random sample of 250 pieces taken from a large batch of components will contain more than 2 defectives if the batch is
   (a) 0·4 per cent defective
   (b) 2 per cent defective.
3. A single sampling inspection scheme states 'Take a random sample of 50 pieces from the batch, if it contains less than 2 defectives, accept the batch, otherwise reject the batch.'

Write down the equation of the operating characteristic for the scheme.

Determine

(a) the Producer's Risk for A.Q.L. = 2 per cent
(b) the Consumer's Risk for L.Q. = 6 per cent.

4. A double sampling Inspection scheme states 'Take a random sample of 30 pieces from the batch, if it contains 0 defectives, accept the batch, if it contains 1 or 2 defectives, take a second sample of 20 pieces. If the total number of defectives in the two samples is less than 3, accept the batch, otherwise reject the batch.'

Write down the equation of the operating characteristic for the scheme.

Determine

(a) the Producer's Risk for A.Q.L. = 2 per cent
(b) the Consumer's Risk for L.Q. = 6 per cent
(c) the A.O.Q. and average sample size ($\bar{n}$) for a series of batches arriving at inspection containing (i) 2 per cent, (ii) 6 per cent of defectives.

5. The inside diameters of a random sample of 100 collars, mass produced to a nominal bore of 7·25 ± 0·04 mm, were measured and are recorded below.

7·260	7·200	7·265	7·280	7·230	7·256	7·248	7·240	7·241	7·271
7·198	7·260	7·215	7·255	7·259	7·294	7·311	7·262	7·287	7·305
7·270	7·255	7·204	7·256	7·238	7·266	7·214	7·237	7·251	7·257
7·275	7·250	7·264	7·239	7·256	7·258	7·290	7·258	7·258	7·268
7·265	7·210	7·255	7·217	7·285	7·258	2·264	7·291	7·234	7·268
7·314	7·258	7·239	7·255	7·258	7·229	7·266	7·235	7·313	7·222
7·255	7·269	7·260	7·283	7·272	7·321	7·256	7·315	7·266	7·280
7·283	7·310	7·273	7·269	7·265	7·261	7·224	7·251	7·257	7·242
7·333	7·262	7·300	7·275	7·220	7·259	7·327	7·313	7·311	7·249
7·219	7·252	7·262	7·259	7·330	7·227	7·237	7·256	7·311	7·245

Use class boundaries of 7·195 mm to $7{\cdot}215^{-}$ mm, 7·215 mm to $7{\cdot}235^{-}$ mm, etc., to form these data into a frequency distribution.

Use this to plot

(a) a frequency histogram
(b) a relative frequency (less than) ogive.

Determine

(c) the median diameter,
(d) the modal class, and
(e) the semi-interquartile range.

Calculate

(f) the mean diameter
(g) the standard deviation of diameter and
(h) the coefficient of variation.

What proportion of the output would you expect to conform to specification?

6. A company manufacturing laminated plastic panels wishes to use a suitable control chart to maintain the quality of its product when produced by a new process.

   The number of minor defects per panel, detected on the first 50 panels produced, were as follows

No. of defects	0	1	2	3	4	5	6
No. of panels	10	21	7	3	5	3	1

   Use these data to set up an appropriate chart.

7. Ball bearings for use within a certain cycle factory are poured straight from the production line into linen bags, each of which holds approximately 250 ball bearings. The contents of the first 30 bags filled after starting up the process were counted and the number of defectives noted, with the following results:

Bag No.	No. of Ball Bearings	No. Defective	Bag No.	No. of Ball Bearings	No. Defective
1	248	1	16	250	5
2	257	1	17	249	1
3	254	4	18	243	2
4	253	4	19	258	5
5	253	2	20	251	4
6	242	4	21	252	6
7	250	1	22	249	5
8	251	1	23	239	3
9	240	4	24	256	1
10	260	2	25	254	0
11	255	2	26	248	6
12	254	1	27	248	2
13	238	6	28	250	5
14	241	0	29	260	0
15	255	1	30	252	2

   Use these data to set up a control chart suitable for maintaining the quality of future output.

8. Nominal 100-watt electric light bulbs are being manufactured by an automatic machine. Ten random samples, each of 4 bulbs, were tested to determine accurately their electrical power consumption, with the following results.

Power–watts–difference from nominal:

Sample No.	1	2	3	4	5
Power difference (watts)	+0·1	−0·1	+0·5	+1·2	−0·7
	+0·8	−0·2	+0·1	−0·8	−0·2
	+0·2	+0·2	0	−0·7	+0·1
	+1·1	+0·7	−0·6	+0·3	+0·4

Sample No.	6	7	8	9	10
Power difference (watts)	−0·3	+0·7	0	+0·5	+1·1
	+0·3	+1·1	+0·7	−0·1	−0·7
	−0·1	−0·2	−0·2	0	+0·4
	−0·1	0	+0·3	+0·2	+0·6

Use these data to set up control charts suitable to detect any change in wattage in the future output of the machine.

9. A chemical is bought by a certain firm in large crystals, packed in cartons which are filled but not weighed. From a consignment of 250 cartons a random sample of 10 cartons was taken and the contents weighed (in grams), with the following results:

Weight g. 518, 492, 504, 521, 489,
514, 510, 508, 512, 497.

Determine 95 per cent confidence limits for the weight of the consignment.

10. To compare two gauges a length of round bar, nominal 7 mm diameter, was measured by 12 people on each gauge, the difference from nominal diameter being recorded.

(a) Initially all 12 used gauge A and then, later, used gauge B, not necessarily in the same order.

Difference from nominal: mm

Gauge A	+0·20	+0·23	+0·07	+0·10	+0·09	−0·01
	−0·06	+0·18	+0·02	+0·07	−0·02	+0·05
Gauge B	+0·20	+0·17	−0·02	+0·06	0	−0·08
	+0·05	+0·07	+0·15	+0·06	+0·11	+0·02

Does the evidence suggest a difference between the readings of the two gauges?

(b) After taking advice, the investigation was repeated, each person measuring on both gauges before passing the bar on to the next person.

Gauge A	+0·21	−0·05	+0·18	+0·03	+0·07	+0·21
B	+0·20	−0·07	+0·16	+0·05	+0·04	+0·18
Gauge A	+0·05	−0·03	+0·10	+0·11	+0·06	−0·02
B	+0·07	0	+0·10	+0·08	+0·07	−0·03

Re-examine your conclusions, reached on the evidence of part (a), and state your new conclusions.

11. Thicknesses of two sheets of steel were each measured (in mm) at 9 random positions, and recorded, as follows:

Sheet A	1·212	1·209	1·214	1·210	1·207
	1·206	1·214	1·211	1·208	
Sheet B	1·201	1·205	1·200	1·209	1·202
	1·201	1·207	1·206	1·204	

Does this evidence suggest that the thickness of one sheet is more variable than the other?

12. In a plant producing liquid detergent 100 random samples are taken each day from the output and their concentrations measured.

The concentrations, relative to an established standard, on two consecutive days, were found to have means of 1·008 and 1·006 with standard deviations of 0·0030 and 0·0025 respectively. Test the hypothesis that the concentration did not differ from day to day.

13. Over a long period of time it has been established that the number of requisitions per hour, on an 8-hour working day, received at a works stores has a mean of 47 and a standard deviation of 6. On a given day the number of requisitions averaged 55 per hour. Test the hypothesis that it was not an unusually busy day.

14. Measurements of the extension of a bar under a given load were, in mm,

0·166	0·165	0·163	0·168	0·164	0·168
0·167	0·169	0·166	0·167		

Do these measurements conflict with the manufacturer's claim that, for this load, the extension is 0·165 mm?

15. An extensive works site has an internal telephone system. The number of telephone failures reported daily are noted and a summary for a given period is as follows.

No. of failures reported per day x	No. of days having x failures
0	98
1	59
2	32
3	8
4	2
5	1
6	0

Test the hypothesis that the number of failures per day follows a Poisson distribution.

Glossary

Acceptable Quality Level (A.Q.L.) Sampling Inspection. A term used to denote a small proportion of defectives in a batch, such that batches containing not more than this proportion of defectives have a high probability of being accepted.
Acceptance Number Sampling Inspection. The maximum number of defectives which a sample may contain if the batch is to be accepted.
Average Outgoing Quality (A.O.Q.) Sampling Inspection. The average proportion defective after inspection, taken over many batches arriving at Inspection containing the same proportion defective. 'Rejected' batches having been 100 per cent inspected, all defectives are removed and the batch size made up with 'good' items.
Average Outgoing Quality Limit (A.O.Q.L.) Sampling Inspection. The highest value that A.O.Q. can take whatever the proportion of defectives present in the batches arriving at Inspection.
Batch (or Lot) A specific quantity of a particular item, assumed to be manufactured under uniform conditions.
Batch Size The number of items in a batch.
Class Boundaries Those values of the variate which mark the end of one class (or group) and the beginning of another when grouping data.
Class Limits The greatest and least values of the variate to be included in a given class (or group) when grouping discrete or 'rounded' continuous data.
Class Mid-Mark Half the sum of the class boundaries for each class.
Consumer's Risk Sampling Inspection. The risk that a batch containing an excessive proportion of defectives will yield a 'good' sample and hence be accepted, when it should have been rejected.
Continuous Data Data which can take any value in a given range. Usually data obtained by measurement.
Data A set of measurements or figures to be analyzed.
Discrete Data Data which can only take specified values and nothing in between. Usually data obtained by counting.
Distribution (Frequency distribution) The relationship between the values of the data and the frequencies with which these values occur.

Double Sampling Sampling Inspection. A system in which the decision to accept or reject a batch may be reached after not more than two samples have been inspected.

Empirical Probability A measure of probability based on observation of the frequency of occurrence of the given event over a series of trials.

Frequency The number of times a given event, size, dimension or characteristic occurs.

Histogram A pictorial representation of a frequency distribution.

Hypothesis A statement or claim concerning a characteristic of a population, to be tested by the data available.

Limiting Quality (L.Q.) Sampling Inspection. A small proportion defective, somewhat larger than A.Q.L., such that batches containing more than this proportion of defectives should have only a small probability of being accepted.

Lot Tolerance Percent Defective (L.T.P.D.) Same as L.Q.

Mean A measure of location of a distribution. Calculated as the sum of the values of the variate divided by the number of values summed.

Median That value of the variate which divides the total frequency, arranged in ascending order, into two equal parts.

Mid-mark See class mid-mark.

Mode That value of the variate which occurs with the greatest frequency.

Ogive A pictorial presentation of a cumulative frequency distribution.

Operating Characteristic Sampling Inspection. For a given scheme a graph of probability of accepting a batch plotted against proportion of defectives in the batch. Also used for the equation of such a curve.

Probability A mathematical measure of the likelihood of an event, defined as follows 'If a trial can result in any one of *n* exhaustive, mutually exclusive, and equally likely events, and *m* of these events are regarded as "successes" the probability that any one trial will result in "success" is m/n'.

Process Capability The limits of the variation in dimension from piece to piece which will inevitably occur due to chance causes. Usually taken to be Mean $\pm$ three standard deviations, i.e., the 'tolerance' within which 99·7 per cent of articles, made by a mass-production process, may be expected to fall.

Producer's Risk Sampling Inspection. The risk that a batch containing an acceptably small proportion of defectives will yield a bad sample, and hence be rejected, when it should have been accepted.

Quartiles Those three values of the variate which divide the total frequency, arranged in ascending order, into four equal parts. The middle quartile coincides with the median.

Range The difference between the greatest and least values in a set of data.

Sample A number of items taken from a batch, lot or population.

Sequential Sampling Sampling Inspection. A system in which pieces are extracted singly and at random from a batch, inspected, and a 'score' calculated. The decision to accept the batch, reject the batch or continue sampling is taken on the value of the 'score' achieved after inspecting each piece.

Single Sampling Sampling Inspection. A system in which the decision to accept or reject a batch is taken after inspecting a single sample containing a pre-determined number of items.

Standard Deviation A measure of the 'spread' of a distribution. The root mean square of the deviations of the observations from their mean.

Standard Error Virtually the same as standard deviation. 'Standard error' is usually used when referring to evidence based on a sample, while standard deviation is used for reference to a population.

Variance The square of the standard deviation.

TABLE A CUMULATIVE BINOMIAL PROBABILITIES

The values tabulated are the probabilities of obtaining *r or less* successes in a series of n trials in which the probability of success in a single trial, p, remains constant.

The values tabulated are

$$\sum_{x=0}^{x=r} {}^{n}C_{x}p^{x}(1-p)^{n-x}$$

	$p =$	0·01	0·02	0·03	0·04	0·05	0·06	0·07	0·08	0·09	0·10
$n = 2$	$r = 0$	0·9801	0·9604	0·9409	0·9216	0·9025	0·8836	0·8649	0·8464	0·8281	0·8100
	1	0·9999	0·9996	0·9991	0·9984	0·9975	0·9964	0·9951	0·9936	0·9919	0·9900
	2	1·0000	1·0000	1·0000	1·0000	1·0000	1·0000	1·0000	1·0000	1·0000	1·0000
$n = 3$	$r = 0$	0·9703	0·9412	0·9127	0·8847	0·8574	0·8306	0·8044	0·7787	0·7536	0·7290
	1	0·9997	0·9988	0·9974	0·9953	0·9928	0·9896	0·9860	0·9818	0·9772	0·9720
	2	1·0000	1·0000	1·0000	0·9999	0·9999	0·9998	0·9996	0·9995	0·9993	0·9990
	3				1·0000	1·0000	1·0000	1·0000	1·0000	1·0000	1·0000
$n = 4$	$r = 0$	0·9606	0·9224	0·8853	0·8493	0·8145	0·7807	0·7480	0·7164	0·6857	0·6561
	1	0·9994	0·9977	0·9948	0·9909	0·9860	0·9801	0·9733	0·9656	0·9570	0·9477
	2	1·0000	1·0000	0·9999	0·9998	0·9995	0·9992	0·9987	0·9981	0·9973	0·9963
	3			1·0000	1·0000	1·0000	1·0000	1·0000	1·0000	0·9999	0·9999
	4									1·0000	1·0000
$n = 5$	$r = 0$	0·9510	0·9039	0·8587	0·8154	0·7738	0·7339	0·6957	0·6591	0·6240	0·5905
	1	0·9990	0·9962	0·9915	0·9852	0·9774	0·9681	0·9575	0·9456	0·9326	0·9185
	2	1·0000	0·9999	0·9997	0·9994	0·9988	0·9980	0·9969	0·9955	0·9936	0·9914
	3		1·0000	1·0000	1·0000	1·0000	0·9999	0·9999	0·9998	0·9997	0·9995
	4						1·0000	1·0000	1·0000	1·0000	1·0000
$n = 10$	$r = 0$	0·9044	0·8171	0·7374	0·6648	0·5987	0·5386	0·4840	0·4344	0·3894	0·3487
	1	0·9957	0·9838	0·9655	0·9418	0·9139	0·8824	0·8483	0·8121	0·7746	0·7361
	2	0·9999	0·9991	0·9972	0·9938	0·9885	0·9812	0·9716	0·9599	0·9460	0·9298
	3	1·0000	1·0000	0·9998	0·9996	0·9990	0·9980	0·9964	0·9942	0·9912	0·9872
	4			1·0000	1·0000	0·9999	0·9998	0·9997	0·9994	0·9990	0·9984
	5					1·0000	1·0000	1·0000	1·0000	0·9999	0·9998
	6									1·0000	1·0000

	$p =$	0·01	0·02	0·03	0·04	0·05	0·06	0·07	0·08	0·09	0·10
$n = 15$	$r = 0$	0·8600	0·7386	0·6332	0·5421	0·4633	0·3953	0·3367	0·2863	0·2430	0·2059
	1	0·9904	0·9647	0·9270	0·8809	0·8290	0·7738	0·7168	0·6597	0·6035	0·5490
	2	0·9996	0·9970	0·9906	0·9797	0·9638	0·9429	0·9171	0·8870	0·8531	0·8159
	3	1·0000	0·9998	0·9992	0·9976	0·9945	0·9896	0·9825	0·9727	0·9600	0·9444
	4		1·0000	0·9999	0·9998	0·9994	0·9986	0·9972	0·9950	0·9918	0·9873
	5			1·0000	1·0000	0·9999	0·9998	0·9997	0·9993	0·9987	0·9978
	6					1·0000	1·0000	1·0000	0·9999	0·9998	0·9997
	7								1·0000	1·0000	1·0000
$n = 20$	$r = 0$	0·8179	0·6676	0·5438	0·4420	0·3585	0·2901	0·2342	0·1887	0·1516	0·1216
	1	0·9831	0·9401	0·8802	0·8103	0·7358	0·6604	0·5868	0·5168	0·4516	0·3917
	2	0·9990	0·9929	0·9790	0·9561	0·9245	0·8850	0·8390	0·7879	0·7334	0·6769
	3	1·0000	0·9994	0·9973	0·9926	0·9841	0·9710	0·9529	0·9294	0·9007	0·8670
	4		1·0000	0·9997	0·9990	0·9974	0·9944	0·9893	0·9816	0·9710	0·9568
	5			1·0000	0·9999	0·9997	0·9991	0·9981	0·9962	0·9932	0·9887
	6				1·0000	1·0000	0·9999	0·9997	0·9994	0·9987	0·9976
	7						1·0000	1·0000	0·9999	0·9998	0·9996
	8								1·0000	1·0000	0·9999
	9										1·0000
$n = 25$	$r = 0$	0·7778	0·6035	0·4670	0·3604	0·2774	0·2129	0·1630	0·1244	0·0946	0·0718
	1	0·9742	0·9114	0·8280	0·7358	0·6424	0·5527	0·4696	0·3947	0·3286	0·2712
	2	0·9980	0·9868	0·9620	0·9235	0·8729	0·8129	0·7466	0·6768	0·6063	0·5371
	3	0·9999	0·9986	0·9938	0·9835	0·9659	0·9402	0·9064	0·8649	0·8168	0·7636
	4	1·0000	0·9999	0·9992	0·9972	0·9928	0·9850	0·9726	0·9548	0·9314	0·9020
	5		1·0000	0·9999	0·9996	0·9988	0·9969	0·9935	0·9877	0·9790	0·9666
	6			1·0000	1·0000	0·9998	0·9995	0·9987	0·9972	0·9946	0·9905
	7					1·0000	0·9999	0·9998	0·9995	0·9988	0·9977
	8						1·0000	1·0000	0·9999	0·9998	0·9995
	9								1·0000	1·0000	0·9999
	10										1·0000

	p =	0·01	0·02	0·03	0·04	0·05	0·06	0·07	0·08	0·09	0·10
$n = 30$ $r = 0$		0·7397	0·5455	0·4010	0·2938	0·2146	0·1562	0·1134	0·0820	0·0590	0·0424
1		0·9638	0·8794	0·7731	0·6612	0·5535	0·4555	0·3694	0·2958	0·2343	0·1837
2		0·9967	0·9783	0·9399	0·8831	0·8122	0·7324	0·6487	0·5654	0·4855	0·4114
3		0·9998	0·9971	0·9881	0·9694	0·9392	0·8974	0·8450	0·7842	0·7175	0·6474
4		1·0000	0·9997	0·9982	0·9937	0·9844	0·9685	0·9447	0·9126	0·8723	0·8245
5			1·0000	0·9998	0·9989	0·9967	0·9920	0·9838	0·9707	0·9519	0·9268
6				1·0000	0·9998	0·9994	0·9983	0·9960	0·9918	0·9848	0·9742
7					1·0000	0·9999	0·9997	0·9992	0·9980	0·9959	0·9922
8						1·0000	1·0000	0·9998	0·9996	0·9990	0·9980
9								1·0000	0·9999	0·9998	0·9995
10									1·0000	1·0000	0·9999
11											1·0000
$n = 40$ $r = 0$		0·6690	0·4457	0·2957	0·1954	0·1285	0·0842	0·0549	0·0356	0·0230	0·0148
1		0·9393	0·8093	0·6615	0·5210	0·3991	0·2990	0·2201	0·1594	0·1140	0·0805
2		0·9925	0·9543	0·8822	0·7855	0·6767	0·5665	0·4625	0·3694	0·2894	0·2228
3		0·9993	0·9918	0·9686	0·9252	0·8618	0·7827	0·6937	0·6007	0·5092	0·4231
4		1·0000	0·9988	0·9933	0·9790	0·9520	0·9104	0·8546	0·7868	0·7103	0·6290
5			0·9999	0·9988	0·9951	0·9861	0·9691	0·9418	0·9033	0·8535	0·7937
6			1·0000	0·9998	0·9990	0·9966	0·9904	0·9801	0·9624	0·9361	0·9005
7				1·0000	0·9998	0·9993	0·9977	0·9942	0·9873	0·9758	0·9581
8					1·0000	0·9999	0·9995	0·9985	0·9963	0·9919	0·9845
9						1·0000	0·9999	0·9997	0·9990	0·9976	0·9949
10							1·0000	0·9999	0·9998	0·9994	0·9985
11								1·0000	1·0000	0·9999	0·9996
12										1·0000	0·9999
13											1·0000

	$p =$	0·01	0·02	0·03	0·04	0·05	0·06	0·07	0·08	0·09	0·10
$n = 50$	$r = 0$	0·6050	0·3642	0·2181	0·1299	0·0769	0·0453	0·0266	0·0155	0·0090	0·0052
	1	0·9106	0·7358	0·5553	0·4005	0·2794	0·1900	0·1265	0·0827	0·0532	0·0338
	2	0·9862	0·9216	0·8108	0·6767	0·5405	0·4162	0·3108	0·2260	0·1605	0·1117
	3	0·9984	0·9822	0·9372	0·8609	0·7604	0·6473	0·5327	0·4253	0·3303	0·2503
	4	0·9998	0·9968	0·9832	0·9510	0·8964	0·8206	0·7290	0·6290	0·5276	0·4312
	5	1·0000	0·9995	0·9963	0·9856	0·9622	0·9224	0·8650	0·7919	0·7072	0·6161
	6		0·9999	0·9993	0·9964	0·9882	0·9711	0·9417	0·8981	0·8404	0·7702
	7		1·0000	0·9999	0·9992	0·9968	0·9906	0·9780	0·9562	0·9232	0·8778
	8			1·0000	0·9998	0·9992	0·9973	0·9927	0·9833	0·9672	0·9421
	9				1·0000	0·9998	0·9993	0·9978	0·9944	0·9875	0·9755
	10					1·0000	0·9998	0·9994	0·9983	0·9957	0·9906
	12						1·0000	1·0000	0·9999	0·9996	0·9990
	14								1·0000	1·0000	0·9999
	16										1·0000

	$p =$	0·15	0·20	0·25	0·30	0·35	0·40	0·45	0·50
$n = 2$	$r = 0$	0·7225	0·6400	0·5625	0·4900	0·4225	0·3600	0·3025	0·2500
	1	0·9775	0·9600	0·9375	0·9100	0·8775	0·8400	0·7975	0·7500
	2	1·0000	1·0000	1·0000	1·0000	1·0000	1·0000	1·0000	1·0000
$n = 3$	$r = 0$	0·6141	0·5120	0·4219	0·3430	0·2746	0·2160	0·1664	0·1250
	1	0·9392	0·8960	0·8437	0·7840	0·7182	0·6480	0·5748	0·5000
	2	0·9966	0·9920	0·9844	0·9730	0·9571	0·9360	0·9089	0·8750
	3	1·0000	1·0000	1·0000	1·0000	1·0000	1·0000	1·0000	1·0000
$n = 4$	$r = 0$	0·5220	0·4096	0·3164	0·2401	0·1785	0·1296	0·0915	0·0625
	1	0·8905	0·8192	0·7383	0·6517	0·5630	0·4752	0·3910	0·3125
	2	0·9880	0·9728	0·9492	0·9163	0·8735	0·8208	0·7585	0·6875
	3	0·9995	0·9984	0·9961	0·9919	0·9850	0·9744	0·9590	0·9375
	4	1·0000	1·0000	1·0000	1·0000	1·0000	1·0000	1·0000	1·0000
$n = 5$	$r = 0$	0·4437	0·3277	0·2373	0·1681	0·1160	0·0778	0·0503	0·0312
	1	0·8352	0·7373	0·6328	0·5282	0·4284	0·3370	0·2562	0·1875
	2	0·9734	0·9421	0·8965	0·8369	0·7648	0·6826	0·5931	0·5000
	3	0·9978	0·9933	0·9844	0·9692	0·9460	0·9130	0·8688	0·8125
	4	0·9999	0·9997	0·9990	0·9976	0·9947	0·9898	0·9815	0·9687
	5	1·0000	1·0000	1·0000	1·0000	1·0000	1·0000	1·0000	1·0000
$n = 10$	$r = 0$	0·1969	0·1074	0·0563	0·0282	0·0135	0·0060	0·0025	0·0010
	1	0·5443	0·3758	0·2440	0·1493	0·0860	0·0464	0·0232	0·0107
	2	0·8202	0·6778	0·5256	0·3828	0·2616	0·1673	0·0996	0·0547
	3	0·9500	0·8791	0·7759	0·6496	0·5138	0·3823	0·2660	0·1719
	4	0·9901	0·9672	0·9219	0·8497	0·7515	0·6331	0·5044	0·3770
	5	0·9986	0·9936	0·9803	0·9526	0·9051	0·8338	0·7384	0·6230
	6	0·9999	0·9991	0·9965	0·9894	0·9740	0·9452	0·8980	0·8281
	7	1·0000	0·9999	0·9996	0·9984	0·9952	0·9877	0·9726	0·9453
	8		1·0000	1·0000	0·9999	0·9995	0·9983	0·9955	0·9892
	9				1·0000	1·0000	0·9999	0·9996	0·9990
	10						1·0000	1·0000	1·0000

	$p =$	0·15	0·20	0·25	0·30	0·35	0·40	0·45	0·50
$n = 15$	$r = 0$	0·0874	0·0352	0·0134	0·0047	0·0016	0·0005	0·0001	
	1	0·3186	0·1671	0·0802	0·0353	0·0142	0·0052	0·0017	0·0005
	2	0·6042	0·3980	0·2361	0·1268	0·0617	0·0271	0·0106	0·0037
	3	0·8226	0·6482	0·4613	0·2969	0·1727	0·0905	0·0424	0·0176
	4	0·9383	0·8358	0·6865	0·5155	0·3519	0·2173	0·1204	0·0592
	5	0·9832	0·9389	0·8516	0·7216	0·5643	0·4032	0·2608	0·1509
	6	0·9964	0·9819	0·9434	0·8688	0·7548	0·6098	0·4522	0·3036
	7	0·9994	0·9958	0·9827	0·9500	0·8868	0·7869	0·6535	0·5000
	8	0·9999	0·9992	0·9958	0·9848	0·9578	0·9050	0·8182	0·6964
	9	1·0000	0·9999	0·9992	0·9963	0·9876	0·9662	0·9231	0·8491
	10		1·0000	0·9999	0·9993	0·9972	0·9907	0·9745	0·9408
	11			1·0000	0·9999	0·9995	0·9981	0·9937	0·9824
	12				1·0000	0·9999	0·9997	0·9989	0·9963
	13					1·0000	1·0000	0·9999	0·9995
	14							1·0000	1·0000
$n = 20$	$r = 0$	0·0388	0·0115	0·0032	0·0008	0·0002			
	1	0·1756	0·0692	0·0243	0·0076	0·0021	0·0005	0·0001	
	2	0·4049	0·2061	0·0913	0·0355	0·0121	0·0036	0·0009	0·0002
	3	0·6477	0·4114	0·2212	0·1071	0·0444	0·0160	0·0049	0·0013
	4	0·8298	0·6296	0·4148	0·2375	0·1182	0·0510	0·0189	0·0059
	5	0·9327	0·8042	0·6172	0·4164	0·2454	0·1256	0·0553	0·0207
	6	0·9781	0·9133	0·7858	0·6080	0·4166	0·2500	0·1299	0·0577
	7	0·9941	0·9678	0·8982	0·7723	0·6010	0·4159	0·2520	0·1316
	8	0·9987	0·9900	0·9591	0·8867	0·7624	0·5956	0·4143	0·2517
	9	0·9998	0·9974	0·9861	0·9520	0·8782	0·7553	0·5914	0·4119
	10	1·0000	0·9994	0·9960	0·9828	0·9468	0·8725	0·7507	0·5881
	12		1·0000	0·9998	0·9987	0·9940	0·9790	0·9420	0·8684
	14			1·0000	1·0000	0·9997	0·9984	0·9936	0·9793
	16					1·0000	1·0000	0·9997	0·9987
	18							1·0000	1·0000

	p =	0·15	0·20	0·25	0·30	0·35	0·40	0·45	0·50
n = 25	r = 0	0·0172	0·0038	0·0008	0·0001				
	1	0·0931	0·0274	0·0070	0·0016	0·0003			
	2	0·2537	0·0982	0·0321	0·0090	0·0021	0·0004	0·0001	
	3	0·4711	0·2340	0·0962	0·0332	0·0097	0·0024	0·0005	0·0001
	4	0·6821	0·4207	0·2137	0·0905	0·0320	0·0095	0·0023	0·0005
	5	0·8385	0·6167	0·3783	0·1935	0·0826	0·0294	0·0086	0·0020
	6	0·9305	0·7800	0·5611	0·3406	0·1734	0·0736	0·0256	0·0073
	7	0·9745	0·8909	0·7265	0·5118	0·3061	0·1536	0·0638	0·0216
	8	0·9920	0·9532	0·8506	0·6769	0·4668	0·2735	0·1340	0·0539
	9	0·9978	0·9826	0·9287	0·8106	0·6303	0·4246	0·2424	0·1148
	10	0·9995	0·9944	0·9703	0·9022	0·7712	0·5858	0·3843	0·2122
	12	1·0000	0·9996	0·9966	0·9825	0·9396	0·8462	0·6937	0·5000
	14		1·0000	0·9998	0·9982	0·9907	0·9656	0·9040	0·7878
	16			1·0000	0·9999	0·9992	0·9957	0·9826	0·9461
	18				1·0000	1·0000	0·9997	0·9984	0·9927
	20						1·0000	0·9999	0·9995
	22							1·0000	1·0000
n = 30	r = 0	0·0076	0·0012	0·0001					
	1	0·0480	0·0105	0·0020	0·0003				
	2	0·1514	0·0442	0·0106	0·0021	0·0003			
	3	0·3216	0·1227	0·0374	0·0093	0·0019	0·0003		
	4	0·5245	0·2552	0·0979	0·0302	0·0075	0·0015	0·0002	
	5	0·7106	0·4275	0·2026	0·0766	0·0232	0·0057	0·0011	0·0002
	6	0·8474	0·6070	0·3480	0·1595	0·0586	0·0172	0·0040	0·0007
	7	0·9302	0·7608	0·5143	0·2814	0·1238	0·0435	0·0121	0·0026
	8	0·9722	0·8713	0·6736	0·4315	0·2247	0·0940	0·0312	0·0081
	9	0·9903	0·9389	0·8034	0·5888	0·3575	0·1763	0·0694	0·0214
	10	0·9970	0·9744	0·8943	0·7304	0·5078	0·2915	0·1350	0·0494
	12	0·9998	0·9969	0·9784	0·9155	0·7802	0·5785	0·3592	0·1808
	14	1·0000	0·9998	0·9972	0·9831	0·9348	0·8246	0·6448	0·4278
	16		1·0000	0·9998	0·9979	0·9876	0·9519	0·8644	0·7077
	18			1·0000	0·9998	0·9986	0·9917	0·9666	0·8998
	20				1·0000	0·9999	0·9991	0·9950	0·9786
	22					1·0000	1·0000	0·9996	0·9974
	24							1·0000	0·9998
	26								1·0000

$n = 40$ $r = 0$	0·0015	0·0001						
1	0·0121	0·0015	0·0001					
2	0·0486	0·0079	0·0010	0·0001				
3	0·1302	0·0285	0·0047	0·0006				
4	0·2633	0·0759	0·0160	0·0026	0·0003			
5	0·4325	0·1613	0·0433	0·0086	0·0013	0·0001		
6	0·6066	0·2859	0·0962	0·0238	0·0044	0·0006		
7	0·7559	0·4371	0·1820	0·0553	0·0124	0·0020	0·0002	
8	0·8646	0·5931	0·2998	0·1110	0·0302	0·0061	0·0009	0·0001
9	0·9328	0·7318	0·4395	0·1959	0·0644	0·0156	0·0027	0·0003
10	0·9701	0·8392	0·5839	0·3087	0·1215	0·0352	0·0074	0·0011
12	0·9957	0·9568	0·8209	0·5772	0·3143	0·1285	0·0386	0·0083
14	0·9996	0·9921	0·9456	0·8074	0·5721	0·3174	0·1326	0·0403
16	1·0000	0·9990	0·9884	0·9367	0·7978	0·5681	0·3185	0·1341
18		0·9999	0·9983	0·9852	0·9301	0·7911	0·5650	0·3179
20		1·0000	0·9998	0·9976	0·9827	0·9256	0·7870	0·5627
25			1·0000	1·0000	0·9999	0·9988	0·9914	0·9596
30					1·0000	1·0000	1·0000	0·9997
35								1·0000
$n = 50$ $r = 0$	0·0003							
1	0·0029	0·0002						
2	0·0142	0·0013	0·0001					
3	0·0460	0·0056	0·0005					
4	0·1121	0·0185	0·0021	0·0002				
5	0·2194	0·0480	0·0070	0·0007				
6	0·3613	0·1034	0·0194	0·0025	0·0002			
7	0·5188	0·1904	0·0452	0·0073	0·0008	0·0001		
8	0·6681	0·3073	0·0916	0·0183	0·0025	0·0002		
9	0·7911	0·4437	0·1637	0·0402	0·0067	0·0008		
10	0·8801	0·5836	0·2622	0·0788	0·0160	0·0022	0·0002	
12	0·9699	0·8139	0·5110	0·2229	0·0661	0·0132	0·0018	0·0002
14	0·9947	0·9393	0·7481	0·4468	0·1878	0·0540	0·0104	0·0013
16	0·9993	0·9856	0·9017	0·6839	0·3888	0·1561	0·0426	0·0077
18	0·9999	0·9975	0·9713	0·8594	0·6216	0·3356	0·1273	0·0324
20	1·0000	0·9997	0·9937	0·9522	0·8139	0·5610	0·2862	0·1013
25		1·0000	1·0000	0·9991	0·9900	0·9426	0·8034	0·5561
30				1·0000	0·9999	0·9986	0·9884	0·9405
35					1·0000	1·0000	0·9999	0·9987
40							1·0000	1·0000

TABLE B CUMULATIVE POISSON PROBABILITIES

The values tabulated are the probabilities of obtaining *r or less* successes in an interval when the average number of successes per interval is λ.

The values tabulated are

$$\sum_{x=0}^{x=r} \frac{e^{-\lambda}\lambda^x}{x!}$$

	$\lambda =$ 0·1	0·2	0·3	0·4	0·5	0·6	0·7	0·8	0·9	1·0
$r = 0$	0·9048	0·8187	0·7408	0·6703	0·6065	0·5488	0·4966	0·4493	0·4066	0·3679
1	0·9953	0·9825	0·9631	0·9384	0·9098	0·8781	0·8442	0·8088	0·7725	0·7358
2	0·9998	0·9988	0·9964	0·9921	0·9856	0·9769	0·9658	0·9526	0·9371	0·9197
3	1·0000	0·9999	0·9997	0·9992	0·9982	0·9966	0·9942	0·9909	0·9865	0·9810
4		1·0000	1·0000	0·9999	0·9998	0·9996	0·9992	0·9986	0·9976	0·9963
5				1·0000	1·0000	1·0000	0·9999	0·9998	0·9996	0·9994
6							1·0000	1·0000	1·0000	0·9999
7										1·0000

	$\lambda =$ 1·1	1·2	1·3	1·4	1·5	1·6	1·7	1·8	1·9	2·0
$r = 0$	0·3329	0·3012	0·2725	0·2466	0·2231	0·2019	0·1827	0·1653	0·1496	0·1353
1	0·6990	0·6626	0·6268	0·5918	0·5578	0·5249	0·4932	0·4628	0·4337	0·4060
2	0·9004	0·8795	0·8571	0·8335	0·8088	0·7834	0·7572	0·7306	0·7037	0·6767
3	0·9742	0·9662	0·9569	0·9463	0·9344	0·9212	0·9068	0·8913	0·8747	0·8571
4	0·9946	0·9922	0·9893	0·9857	0·9814	0·9763	0·9704	0·9636	0·9559	0·9473
5	0·9990	0·9985	0·9978	0·9968	0·9955	0·9940	0·9920	0·9896	0·9868	0·9834
6	0·9998	0·9997	0·9996	0·9994	0·9991	0·9987	0·9981	0·9974	0·9966	0·9955
7	1·0000	1·0000	0·9999	0·9999	0·9998	0·9997	0·9996	0·9994	0·9992	0·9989
8			1·0000	1·0000	1·0000	1·0000	0·9999	0·9999	0·9998	0·9998
9							1·0000	1·0000	1·0000	1·0000

	$\lambda =$ 2·1	2·2	2·3	2·4	2·5	2·6	2·7	2·8	2·9	3·0
$r = 0$	0·1224	0·1108	0·1002	0·0907	0·0821	0·0743	0·0672	0·0608	0·0550	0·0498
1	0·3796	0·3546	0·3308	0·3084	0·2873	0·2674	0·2487	0·2311	0·2146	0·1991
2	0·6496	0·6227	0·5960	0·5697	0·5438	0·5184	0·4936	0·4694	0·4460	0·4232
3	0·8386	0·8194	0·7993	0·7787	0·7576	0·7360	0·7141	0·6919	0·6696	0·6472
4	0·9379	0·9275	0·9162	0·9041	0·8912	0·8774	0·8629	0·8477	0·8318	0·8153
5	0·9796	0·9751	0·9700	0·9643	0·9580	0·9510	0·9433	0·9349	0·9258	0·9161
6	0·9941	0·9925	0·9906	0·9884	0·9858	0·9828	0·9794	0·9756	0·9713	0·9665
7	0·9985	0·9980	0·9974	0·9967	0·9958	0·9947	0·9934	0·9919	0·9901	0·9881
8	0·9997	0·9995	0·9994	0·9991	0·9988	0·9985	0·9981	0·9976	0·9969	0·9962
9	0·9999	0·9999	0·9998	0·9998	0·9997	0·9996	0·9995	0·9993	0·9991	0·9989
10	1·0000	1·0000	1·0000	1·0000	0·9999	0·9999	0·9999	0·9998	0·9998	0·9997
11					1·0000	1·0000	1·0000	1·0000	0·9999	0·9999

	λ = 3·1	3·2	3·3	3·4	3·5	3·6	3·7	3·8	3·9	4·0
r = 0	0·0450	0·0408	0·0369	0·0334	0·0302	0·0273	0·0247	0·0224	0·0202	0·0183
1	0·1847	0·1712	0·1586	0·1468	0·1359	0·1257	0·1162	0·1074	0·0992	0·0916
2	0·4012	0·3799	0·3594	0·3397	0·3208	0·3027	0·2854	0·2689	0·2531	0·2381
3	0·6248	0·6025	0·5803	0·5584	0·5366	0·5152	0·4942	0·4735	0·4532	0·4335
4	0·7982	0·7806	0·7626	0·7442	0·7254	0·7064	0·6872	0·6678	0·6484	0·6288
5	0·9057	0·8946	0·8829	0·8705	0·8576	0·8441	0·8301	0·8156	0·8006	0·7851
6	0·9612	0·9554	0·9490	0·9421	0·9347	0·9267	0·9182	0·9091	0·8995	0·8893
7	0·9858	0·9832	0·9802	0·9769	0·9733	0·9692	0·9648	0·9599	0·9546	0·9489
8	0·9953	0·9943	0·9931	0·9917	0·9901	0·9883	0·9863	0·9840	0·9815	0·9786
9	0·9986	0·9982	0·9978	0·9973	0·9967	0·9960	0·9952	0·9942	0·9931	0·9919
10	0·9996	0·9995	0·9994	0·9992	0·9990	0·9987	0·9984	0·9981	0·9976	0·9972
11	0·9999	0·9999	0·9998	0·9998	0·9997	0·9996	0·9995	0·9994	0·9993	0·9991
12	1·0000	1·0000	1·0000	0·9999	0·9999	0·9999	0·9999	0·9998	0·9998	0·9997
13				1·0000	1·0000	1·0000	1·0000	1·0000	0·9999	0·9999
14									1·0000	1·0000

	λ = 4·1	4·2	4·3	4·4	4·5	4·6	4·7	4·8	4·9	5·0
r = 0	0·0166	0·0150	0·0136	0·0123	0·0111	0·0100	0·0091	0·0082	0·0074	0·0067
1	0·0845	0·0780	0·0719	0·0663	0·0611	0·0563	0·0518	0·0477	0·0439	0·0404
2	0·2238	0·2102	0·1974	0·1851	0·1736	0·1626	0·1523	0·1425	0·1333	0·1246
3	0·4142	0·3954	0·3772	0·3594	0·3423	0·3257	0·3097	0·2942	0·2793	0·2650
4	0·6093	0·5898	0·5704	0·5512	0·5321	0·5132	0·4946	0·4762	0·4582	0·4405
5	0·7693	0·7531	0·7367	0·7199	0·7029	0·6858	0·6684	0·6510	0·6335	0·6160
6	0·8786	0·8675	0·8559	0·8436	0·8310	0·8180	0·8046	0·7908	0·7766	0·7622
7	0·9427	0·9360	0·9290	0·9214	0·9134	0·9049	0·8960	0·8867	0·8769	0·8666
8	0·9755	0·9721	0·9683	0·9642	0·9597	0·9549	0·9497	0·9442	0·9382	0·9319
9	0·9904	0·9889	0·9871	0·9851	0·9829	0·9805	0·9778	0·9748	0·9716	0·9682
10	0·9966	0·9959	0·9952	0·9943	0·9933	0·9922	0·9910	0·9896	0·9880	0·9863
11	0·9989	0·9986	0·9983	0·9780	0·9976	0·9971	0·9966	0·9960	0·9953	0·9945
12	0·9996	0·9996	0·9995	0·9993	0·9992	0·9990	0·9988	0·9986	0·9983	0·9980
13	0·9999	0·9999	0·9998	0·9998	0·9997	0·9997	0·9996	0·9995	0·9994	0·9993
14	1·0000	1·0000	1·0000	0·9999	0·9999	0·9999	0·9999	0·9998	0·9998	0·9998
15				1·0000	1·0000	1·0000	1·0000	1·0000	0·9999	0·9999
16									1·0000	1·0000

	λ = 5·2	5·4	5·6	5·8	6·0	6·2	6·4	6·6	6·8	7·0
r = 0	0·0055	0·0045	0·0037	0·0030	0·0025	0·0020	0·0017	0·0014	0·0011	0·0009
1	0·0342	0·0289	0·0244	0·0206	0·0174	0·0146	0·0123	0·0103	0·0087	0·0073
2	0·1088	0·0948	0·0824	0·0715	0·0620	0·0536	0·0463	0·0400	0·0344	0·0296
3	0·2381	0·2133	0·1906	0·1700	0·1512	0·1342	0·1189	0·1052	0·0928	0·0818
4	0·4061	0·3733	0·3422	0·3127	0·2850	0·2592	0·2351	0·2127	0·1920	0·1730
5	0·5809	0·5461	0·5119	0·4783	0·4457	0·4141	0·3837	0·3547	0·3270	0·3007
6	0·7324	0·7017	0·6702	0·6384	0·6063	0·5742	0·5423	0·5108	0·4799	0·4497
7	0·8449	0·8216	0·7969	0·7710	0·7440	0·7160	0·6873	0·6581	0·6285	0·5987
8	0·9181	0·9026	0·8867	0·8672	0·8472	0·8259	0·8033	0·7796	0·7548	0·7291
9	0·9603	0·9512	0·9409	0·9292	0·9161	0·9016	0·8858	0·8686	0·8502	0·8305
10	0·9823	0·9775	0·9718	0·9601	0·9574	0·9486	0·9386	0·9274	0·9151	0·9015
11	0·9927	0·9904	0·9875	0·9840	0·9799	0·9750	0·9693	0·9627	0·9552	0·9466
12	0·9972	0·9962	0·9948	0·9932	0·9912	0·9887	0·9857	0·9821	0·9779	0·9730
13	0·9990	0·9986	0·9980	0·9973	0·9964	0·9952	0·9937	0·9920	0·9898	0·9872
14	0·9997	0·9995	0·9993	0·9990	0·9986	0·9981	0·9974	0·9966	0·9956	0·9943
15	0·9999	0·9998	0·9998	0·9996	0·9995	0·9993	0·9990	0·9986	0·9982	0·9976
16	1·0000	0·9999	0·9999	0·9999	0·9998	0·9997	0·9996	0·9995	0·9993	0·9990
17		1·0000	1·0000	1·0000	0·9999	0·9999	0·9999	0·9998	0·9997	0·9996
18					1·0000	1·0000	1·0000	0·9999	0·9999	0·9999
19								1·0000	1·0000	1·0000

	λ = 7·2	7·4	7·6	7·8	8·0	8·5	9·0	9·5	10·0
r = 0	0·0007	0·0006	0·0005	0·0004	0·0003	0·0002	0·0001		
1	0·0061	0·0051	0·0043	0·0036	0·0030	0·0019	0·0012	0·0008	0·0005
2	0·0255	0·0219	0·0188	0·0161	0·0138	0·0093	0·0062	0·0042	0·0028
3	0·0719	0·0632	0·0554	0·0485	0·0424	0·0301	0·0212	0·0148	0·0103
4	0·1555	0·1395	0·1249	0·1117	0·0996	0·0744	0·0550	0·0403	0·0292
5	0·2759	0·2526	0·2307	0·2102	0·1912	0·1496	0·1157	0·0885	0·0671
6	0·4204	0·3920	0·3646	0·3384	0·3134	0·2562	0·2068	0·1649	0·1301
7	0·5689	0·5393	0·5100	0·4812	0·4530	0·3856	0·3239	0·2687	0·2202
8	0·7027	0·6756	0·6482	0·6204	0·5925	0·5231	0·4556	0·3918	0·3328
9	0·8096	0·7877	0·7648	0·7411	0·7166	0·6530	0·5874	0·5218	0·4579
10	0·8867	0·8707	0·8535	0·8352	0·8159	0·7634	0·7060	0·6453	0·5830
11	0·9371	0·9265	0·9148	0·9020	0·8881	0·8487	0·8030	0·7520	0·6968
12	0·9673	0·9609	0·9536	0·9454	0·9362	0·9091	0·8758	0·8364	0·7916
13	0·9841	0·9805	0·9762	0·9714	0·9658	0·9486	0·9261	0·8981	0·8645
14	0·9927	0·9908	0·9886	0·9859	0·9827	0·9726	0·9585	0·9400	0·9165
15	0·9968	0·9959	0·9948	0·9934	0·9918	0·9862	0·9780	0·9665	0·9512
16	0·9987	0·9983	0·9978	0·9971	0·9963	0·9934	0·9889	0·9823	0·9730
17	0·9995	0·9993	0·9990	0·9988	0·9984	0·9970	0·9947	0·9911	0·9857
18	0·9998	0·9997	0·9996	0·9995	0·9993	0·9987	0·9976	0·9957	0·9928
19	0·9999	0·9999	0·9999	0·9998	0·9997	0·9995	0·9989	0·9980	0·9965
20	1·0000	1·0000	1·0000	0·9999	0·9999	0·9998	0·9996	0·9991	0·9984
21				1·0000	1·0000	0·9999	0·9998	0·9996	0·9993
22						1·0000	0·9999	0·9998	0·9997
23							1·0000	0·9999	0·9999
24								1·0000	1·0000

TABLE C AREA UNDER THE NORMAL CURVE

The area is measured from the mean, $Z = 0$, to any ordinate Z.

Z	0·00	0·01	0·02	0·03	0·04	0·05	0·06	0·07	0·08	0·09
0·0	0·0000	0·0040	0·0080	0·0120	0·0159	0·0199	0·0239	0·0279	0·0319	0·0359
0·1	0·0398	0·0438	0·0478	0·0517	0·0557	0·0596	0·0636	0·0675	0·0714	0·0753
0·2	0·0793	0·0832	0·0871	0·0910	0·0948	0·0987	0·1026	0·1064	0·1103	0·1141
0·3	0·1179	0·1217	0·1255	0·1293	0·1331	0·1368	0·1406	0·1443	0·1480	0·1517
0·4	0·1554	0·1591	0·1628	0·1664	0·1700	0·1736	0·1772	0·1808	0·1844	0·1879
0·5	0·1915	0·1950	0·1985	0·2019	0·2054	0·2088	0·2123	0·2157	0·2190	0·2224
0·6	0·2257	0·2291	0·2324	0·2357	0·2389	0·2422	0·2454	0·2486	0·2518	0·2549
0·7	0·2580	0·2611	0·2642	0·2673	0·2704	0·2734	0·2764	0·2794	0·2823	0·2852
0·8	0·2881	0·2910	0·2939	0·2967	0·2995	0·3023	0·3051	0·3078	0·3106	0·3133
0·9	0·3159	0·3186	0·3212	0·3238	0·3264	0·3289	0·3315	0·3340	0·3365	0·3389
1·0	0·3413	0·3438	0·3461	0·3485	0·3508	0·3531	0·3554	0·3577	0·3599	0·3621
1·1	0·3643	0·3665	0·3686	0·3708	0·3729	0·3749	0·3770	0·3790	0·3810	0·3830
1·2	0·3849	0·3869	0·3888	0·3907	0·3925	0·3944	0·3962	0·3980	0·3997	0·4015
1·3	0·4032	0·4049	0·4066	0·4082	0·4099	0·4115	0·4131	0·4147	0·4162	0·4177
1·4	0·4192	0·4207	0·4222	0·4236	0·4251	0·4265	0·4279	0·4292	0·4306	0·4319
1·5	0·4332	0·4345	0·4357	0·4370	0·4382	0·4394	0·4406	0·4418	0·4430	0·4441
1·6	0·4452	0·4463	0·4474	0·4485	0·4495	0·4505	0·4515	0·4525	0·4535	0·4545
1·7	0·4554	0·4564	0·4573	0·4582	0·4591	0·4599	0·4608	0·4616	0·4625	0·4633
1·8	0·4641	0·4649	0·4656	0·4664	0·4671	0·4678	0·4686	0·4693	0·4699	0·4706
1·9	0·4713	0·4719	0·4726	0·4732	0·4738	0·4744	0·4750	0·4756	0·4762	0·4767
2·0	0·4772	0·4778	0·4783	0·4788	0·4793	0·4798	0·4803	0·4808	0·4812	0·4817
2·1	0·4821	0·4826	0·4830	0·4834	0·4838	0·4842	0·4846	0·4850	0·4854	0·4857
2·2	0·4861	0·4865	0·4868	0·4871	0·4875	0·4878	0·4881	0·4884	0·4887	0·4890
2·3	0·4893	0·4896	0·4898	0·4901	0·4904	0·4906	0·4909	0·4911	0·4913	0·4916
2·4	0·4918	0·4920	0·4922	0·4925	0·4927	0·4929	0·4931	0·4932	0·4934	0·4936
2·5	0·4938	0·4940	0·4941	0·4943	0·4945	0·4946	0·4948	0·4949	0·4951	0·4952
2·6	0·4953	0·4955	0·4956	0·4957	0·4959	0·4960	0·4961	0·4962	0·4963	0·4964
2·7	0·4965	0·4966	0·4967	0·4968	0·4969	0·4970	0·4971	0·4972	0·4973	0·4974
2·8	0·4974	0·4975	0·4976	0·4977	0·4977	0·4978	0·4979	0·4980	0·4980	0·4981
2·9	0·4981	0·4982	0·4983	0·4983	0·4984	0·4984	0·4985	0·4985	0·4986	0·4986
3·0	0·49865	0·4987	0·4987	0·4988	0·4988	0·4989	0·4989	0·4989	0·4990	0·4990
3·1	0·49903	0·4991	0·4991	0·4991	0·4992	0·4992	0·4992	0·4992	0·4993	0·4993

Weatherburn: *A First Course in Mathematical Statistics.* Cambridge University Press 1962.

TABLE D PERCENTAGE POINTS OF THE t-DISTRIBUTION

2Q%	5	1	0·1
Q%	2·5	0·5	0·05
ν = 1	12·71	63·66	636·6
2	4·30	9·92	31·60
3	3·18	5·84	12·92
4	2·78	4·60	8·61
5	2·57	4·03	6·87
6	2·45	3·71	5·96
7	2·36	3·50	5·41
8	2·31	3·36	5·04
9	2·26	3·25	4·78
10	2·23	3·17	4·59
12	2·18	3·05	4·32
15	2·13	2·95	4·07
20	2·09	2·85	3·85
24	2·06	2·80	3·75
30	2·04	2·75	3·65
40	2·02	2·70	3·55
60	2·00	2·66	3·46
120	1·98	2·62	3·37
∞	1·96	2·58	3·29

← area under curve beyond listed values of t

degrees of freedom (ν) ↓

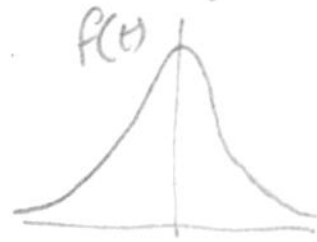

Lindley and Miller: *Cambridge Elementary Statistical Tables,* Cambridge University Press 1961.

TABLE E 5 PER CENT POINTS OF THE F-DISTRIBUTION

$\nu_1 =$	1	2	3	4	5	6	7	8	10	12	24	∞
$\nu_2 =$ 1	161·4	199·5	215·7	224·6	230·2	234·0	236·8	238·9	241·9	243·9	249·0	254·3
2	18·5	19·0	19·2	19·2	19·3	19·3	19·4	19·4	19·4	19·4	19·5	19·5
3	10·13	9·55	9·28	9·12	9·01	8·94	8·89	8·85	8·79	8·74	8·64	8·53
4	7·71	6·94	6·59	6·39	6·26	6·16	6·09	6·04	5·96	5·91	5·77	5·63
5	6·61	5·79	5·41	5·19	5·05	4·95	4·88	4·82	4·74	4·68	4·53	4·36
6	5·99	5·14	4·76	4·53	4·39	4·28	4·21	4·15	4·06	4·00	3·84	3·67
7	5·59	4·74	4·35	4·12	3·97	3·87	3·79	3·73	3·64	3·57	3·41	3·23
8	5·32	4·46	4·07	3·84	3·69	3·58	3·50	3·44	3·35	3·28	3·12	2·93
9	5·12	4·26	3·86	3·63	3·48	3·37	3·29	3·23	3·14	3·07	2·90	2·71
10	4·96	4·10	3·71	3·48	3·33	3·22	3·14	3·07	2·98	2·91	2·74	2·54
11	4·84	3·98	3·59	3·36	3·20	3·09	3·01	2·95	2·85	2·79	2·61	2·40
12	4·75	3·89	3·49	3·26	3·11	3·00	2·91	2·85	2·75	2·69	2·51	2·30
13	4·67	3·81	3·41	3·18	3·03	2·92	2·83	2·77	2·67	2·60	2·42	2·21
14	4·60	3·74	3·34	3·11	2·96	2·85	2·76	2·70	2·60	2·53	2·35	2·13
15	4·54	3·68	3·29	3·06	2·90	2·79	2·71	2·64	2·54	2·48	2·29	2·07
16	4·49	3·63	3·24	3·01	2·85	2·74	2·66	2·59	2·49	2·42	2·24	2·01
17	4·45	3·59	3·20	2·96	2·81	2·70	2·61	2·55	2·45	2·38	2·19	1·96
18	4·41	3·55	3·16	2·93	2·77	2·66	2·58	2·51	2·41	2·34	2·15	1·92
19	4·38	3·52	3·13	2·90	2·74	2·63	2·54	2·48	2·38	2·31	2·11	1·88
20	4·35	3·49	3·10	2·87	2·71	2·60	2·51	2·45	2·35	2·28	2·08	1·84
21	4·32	3·47	3·07	2·84	2·68	2·57	2·49	2·42	2·32	2·25	2·05	1·81
22	4·30	3·44	3·05	2·82	2·66	2·55	2·46	2·40	2·30	2·23	2·03	1·78
23	4·28	3·42	3·03	2·80	2·64	2·53	2·44	2·37	2·27	2·20	2·00	1·76
24	4·26	3·40	3·01	2·78	2·62	2·51	2·42	2·36	2·25	2·18	1·98	1·73
25	4·24	3·39	2·99	2·76	2·60	2·49	2·40	2·34	2·24	2·16	1·96	1·71
26	4·23	3·37	2·98	2·74	2·59	2·47	2·39	2·32	2·22	2·15	1·95	1·69
27	4·21	3·35	2·96	2·73	2·57	2·46	2·37	2·31	2·20	2·13	1·93	1·67
28	4·20	3·34	2·95	2·71	2·56	2·45	2·36	2·29	2·19	2·12	1·91	1·65
29	4·18	3·33	2·93	2·70	2·55	2·43	2·35	2·28	2·18	2·10	1·90	1·64
30	4·17	3·32	2·92	2·69	2·53	2·42	2·33	2·27	2·16	2·09	1·89	1·62
32	4·15	3·29	2·90	2·67	2·51	2·40	2·31	2·24	2·14	2·07	1·86	1·59
34	4·13	3·28	2·88	2·65	2·49	2·38	2·29	2·23	2·12	2·05	1·84	1·57
36	4·11	3·26	2·87	2·63	2·48	2·36	2·28	2·21	2·11	2·03	1·82	1·55
38	4·10	3·24	2·85	2·62	2·46	2·35	2·26	2·19	2·09	2·02	1·81	1·53
40	4·08	3·23	2·84	2·61	2·45	2·34	2·25	2·18	2·08	2·00	1·79	1·51
60	4·00	3·15	2·76	2·53	2·37	2·25	2·17	2·10	1·99	1·92	1·70	1·39
120	3·92	3·07	2·68	2·45	2·29	2·18	2·09	2·02	1·91	1·83	1·61	1·25
∞	3·84	3·00	2·60	2·37	2·21	2·10	2·01	1·94	1·83	1·75	1·52	1·00

Lindley and Miller: *Cambridge Elementary Statistical Tables,* Cambridge University Press 1961.

TABLE F PERCENTAGE POINTS OF THE χ^2 DISTRIBUTION

P%	95	5	1	0·1
ν = 1	0·00393	3·84	6·63	10·83
2	0·103	5·99	9·21	13·81
3	0·352	7·81	11·34	16·27
4	0·711	9·49	13·28	18·47
5	1·15	11·07	15·09	20·52
6	1·64	12·59	16·81	22·46
7	2·17	14·07	18·48	24·32
8	2·73	15·51	20·09	26·12
9	3·33	16·92	21·67	27·88
10	3·94	18·31	23·21	29·59
11	4·57	19·68	24·73	31·26
12	5·23	21·03	26·22	32·91
13	5·89	22·36	27·69	34·53
14	6·57	23·68	29·14	36·12
15	7·26	25·00	30·58	37·70
16	7·96	26·30	32·00	39·25
17	8·67	27·59	33·41	40·79
18	9·39	28·87	34·81	42·31
19	10·12	30·14	36·19	43·82
20	10·85	31·41	37·57	45·31
21	11·59	32·67	38·93	46·80
22	12·34	33·92	40·29	48·27
23	13·09	35·17	41·64	49·73
24	13·85	36·42	42·98	51·18
25	14·61	37·65	44·31	52·62
26	15·38	38·89	45·64	54·05
27	16·15	40·11	46·96	55·48
28	16·93	41·34	48·28	56·89
29	17·71	42·56	49·59	58·30
30	18·49	43·77	50·89	59·70
40	26·51	55·76	63·69	73·40
50	34·76	67·50	76·15	86·66
60	43·19	79·08	88·38	99·61
70	51·74	90·53	100·4	112·3
80	60·39	101·9	112·3	124·8
90	69·13	113·1	124·1	137·2
100	77·93	124·3	135·8	149·4

Based on Lindley and Miller: *Cambridge Elementary Statistical Tables,* Cambridge University Press 1961.

TABLE G

1. *Factor to Estimate Population Standard Deviation* (σ) *from Sample Mean Range* ($\bar{w}$)

$$\text{Estimate of } \sigma = \bar{w} \div d_n \tag{G.1}$$

Sample size n	2	3	4	5	6
d_n	1·13	1·69	2·06	2·33	2·53

2. *Control Chart for Sample Means: 'Warning' and 'Action' Limit Factors*

If $\bar{X}$ = process mean (mean of sample means) and $\bar{w}$ = sample mean range, then

$\frac{1}{40}$ 'Warning' limits for sample mean are $\bar{X} \pm A'_{0 \cdot 025}\bar{w}$ (G.2)

$\frac{1}{1000}$ 'Action' limits for sample mean are $\bar{X} \pm A'_{0\cdot001}\bar{w}$ (G.3)

Sample Size n	2	3	4	5	6
$A'_{0\cdot025}$	1·23	0·67	0·48	0·37	0·32
$A'_{0\cdot001}$	1·94	1·05	0·75	0·59	0·50

3. *Control Chart for Sample Ranges:* $\frac{1}{40}$ *Limit Factors*

The range of 1 sample in 40 may be expected to exceed $D'_{0\cdot975}\bar{w}$ and 1 in 40 may be expected to be less than $D'_{0\cdot025}\bar{w}$, i.e.,

$$\tfrac{1}{40} \text{ Sample range Upper Limit} = D'_{0\cdot975}\bar{w} \qquad \text{(G.4)}$$

$$\tfrac{1}{40} \text{ Sample range Lower Limit} = D'_{0\cdot025}\bar{w} \qquad \text{(G.5)}$$

Sample size	2	3	4	5	6
$D'_{0\cdot975}$	2·81	2·17	1·93	1·81	1·72
$D'_{0\cdot025}$	0·04	0·18	0·29	0·37	0·42

4. *Relative Precision Index*

$$\text{R.P.I.} = \frac{\text{Specification Tolerance}}{\text{Sample Average Range}}$$

	Relative Precision		
Class	Low	Medium	High
Sample size n	Relative Precision Index		
2	less than 6·0	6·0 to 7·0	greater than 7·0
3	less than 4·0	4·0 to 5·0	greater than 5·0
4	less than 3·0	3·0 to 4·0	greater than 4·0
5 and 6	less than 2·5	2·5 to 3·5	greater than 3·5

Answers to problems

Chapter 1

1. (a) 0·02
   (b) 0·98
   (c) 0·0392
   (d) 0·000 4
   (e) 0·9604.
2. (a) 0·1074
   (b) 0·3020
   (c) 0·3222
   (d) 0·999 922
   (e) 302
   (f) 678
   (g) 0
3. (a) 165
   (b) 12
   (c) 236.
4. (a) 6·8 per cent
   (b) 0·1 per cent
   (c) 93·1 per cent.
5. (a) 0·78036 (0·78075)
   (b) 0·20006 (0·19934)
   (c) 0·000 011 485 (0·000 012 96).
6. 83,930
7. (a) 0·30 (b) 0·112 (c) 0·018.
8. (a) 0·72 (b) 0·28 (c) 0·251 (d) 0·029.
9. 0·273.

Chapter 2

1. $P = (1-p)^{19}(1+19p)$

p	0·01	0·02	0·03	0·05	0·07	0·10	0·15	0·20	0·30
P	0·9834	0·9392	0·8803	0·7357	0·5872	0·3918	0·1756	0·0691	0·0075

(a) 1·7 per cent (b) 37 per cent (c) $21\frac{1}{2}$ per cent
A.O.Q.L. = $4\frac{1}{2}$ per cent

2.

	A	B
(a)	5·3%	3·9%
(b)	77%	73%

(c) Scheme B.
(d) Scheme B.

N.B. Both these schemes are poor in that, although they protect the Producer, the risk taken by the Consumer is too great.

Chapter 3

1. $P = P_{(0)}+P_{(1)}+P_{(2)}[P_{(0)}+P_{(1)}+P_{(2)}]+P_{(3)}[P_{(0)}+P_{(1)}]$
$P_{(0)} = (1-p)^{20}$. $P_{(2)} = 190p^2(1-p)^{18}$.
$P_{(1)} = 20p(1-p)^{19}$. $P_{(3)} = 1140p^3(1-p)^{17}$.

p	P	$P_{(0)}$	$P_{(1)}$	$P_{(2)}$	$P_{(3)}$	$\bar{n}$
0·02	0·9977	0·6770	0·2722	0·0528	0·0065	21·2
0·03	0·9931	0·5439	0·3364	0·0988	0·0183	22·3
0·05	0·9539	0·3584	0·3773	0·1887	0·0596	25·0
0·07	0·8662	0·2344	0·3528	0·2523	0·1139	27·3
0·10	0·6574	0·1216	0·2702	0·2852	0·1901	29·4
0·15	0·3113	0·0388	0·1368	0·2296	0·2432	29·5
0·20	0·1115	0·0115	0·0576	0·1368	0·2061	26·9
0·25	0·0334	0·0032	0·0210	0·0665	0·1336	24·0

(a) 4·6 per cent; 25·0 (b) 20·7 per cent; 26·3.

2. $P = P_{(0)}+P_{(1)}+P_{(2)}[P'_{(0)}+P'_{(1)}]+P_{(3)}P'_{(0)}$
where $P_{(0)} = (1-p)^{25}$ $P'_{(0)} = (1-p)^{30}$
$P_{(1)} = {}^{25}C_1(1-p)^{24}p$ $P'_{(1)} = {}^{30}C_1(1-p)^{29}p$
$P_{(2)} = {}^{25}C_2(1-p)^{23}p^2$
$P_{(3)} = {}^{25}C_3(1-p)^{22}p^3$
(a) 1·6 per cent (b) 69 per cent
(c) A.O.Q. 4·1 per cent $\bar{n} = 37{\cdot}6$

3. Accept Batch I after inspecting 24 pieces.
Reject Batch II after inspecting 26 pieces.

4. (a) Reject. Score −26 after 87 pieces inspected.
(b) Accept. Score 350 after 463 pieces inspected.
Handicap = 16. Penalty = 4. Target = 25.

5.

p		0·01	0·15
P	$N = 800$	0·904	0·195
	$N = 100$	0·900	0·181
	Binomial	0·904	0·197

6. (a) Probability of acceptance = $P^2+2PP'(1-P)$.
(b) Average sample size = $20[1+2P(1-P)]$
where $P = P_{(0)}+P_{(1)}$
$= (1-p)^{10}+10p(1-p)^9$.
$P' = P'_{(0)}+P'_{(1)}+P'_{)2)}$
$= (1-p)^{20}+20p(1-p)^{19}+190p^2(1-p)^{18}$.

p	0·02	0·04	0·05	0·07	0·10	0·15	0·20	0·25
Probability of Acceptance	0·9994	0·9918	0·9806	0·9359	0·8050	0·4973	0·2355	0·0931
Average sample size	20·7	22·2	23·2	25·1	27·8	29·9	29·4	27·4

(c) 1·9 per cent; 23·2. (d) 24·8 per cent; 27·5.

Chapter 4

2. (i) 37 per cent approx.
   (ii) 21 per cent approx.
3. Component A: Sample data suggests that the process is capable of meeting the specified tolerance but that adjustment is necessary to reduce the resistances produced by about 0·07 ohm.
   Component B: Sample data indicates that the process is not capable of holding the specified tolerance but the percentage of the resistances produced outside limits would be minimized by reducing the average resistance by about 0·02 ohm.
4. (a) (i) 73 per cent approx.
   (ii) 56 per cent approx.
   (iii) 55 per cent approx.
   (b) (iv) Yes.
   (v) Yes; subject to an overall reduction in diameters of about 0·005 in.
   (vi) No; but percentage of non-conforming work would be minimized by increasing 'aiming' size by about 0·005 in.

Chapter 5

1. Mean 303. Standard deviation 5.
2. (a) Mean 1·03 defectives.
   (b) Standard deviation 0·905 defectives.
   (c) Median 1 defective.
   (d) Mode 1 defective.
3. Mean 0·4975 stoppages.
   Standard deviation 0·721 stoppages.
4. (a) Mean 0·7264 in.
   (b) Standard deviation 0·000 576 in.
   (c) Median 0·72637 in.
5. Mean diameter = 0·735 in.
   Standard deviation of diameter = 0·00525 in.
6. (a) Mean life = 715·5 hours.
   (b) Standard deviation of life = 190·3 hours.
   (c) Coefficient of variation = 26·6 per cent.
   (d) Interquartile range = 293 hours.
7. (a) 1·78 contacts.
   (b) 1·85 contacts.
   (c) 1·12 contacts.
   (d) 2 contacts.
   (e) 1 and 2 contacts (Bimodal).

Chapter 6

1. Hypergeometric.
   Probability of 2 defectives in sample of 8 = 0·0425.
   Mean number of defectives = 0·4.
   Variance of number of defectives = 0·335.

2. (a) Binomial
   - (i) 0·5951
   - (ii) 0·2428
   - (iii) 0·4049
   - (iv) 0·8802
   - (v) 0·0988
   - (vi) 0·0210

   (b) Poisson approximation 0·8781, 0·0988, 0·0231.

   (c)

		Binomial	Poisson
(i)	Mean	3	3
	Variance	2·55	

		Binomial	Poisson
(ii)	Mean	0·6	0·6
	Variance	0·582	

3. Poisson.
   (a) 0·0498 (b) 0·5768 (c) 0·1681.
4. Normal.
   (a) (i) 0·4052 (ii) 0·2877
   (b) (i) 0·4553 in. (ii) 0·4581 in.
5. Normal approximation to Binomial.
   (a) 0·1834 (b) 0·1029 (c) 0·0145
6. Normal.
   (a) 25·6 per cent.
   (b) If mean is raised to 0·2830, non-conforming percentage reduced to 9·6.

Chapter 7

1. Drawing upper limit 0·354 in.
   Upper P.C.L. 0·351 in.
   Lower P.C.L. 0·345 in.
   Drawing lower limit 0·342 in.
2. Drawing upper limit 0·475 in.
   Upper P.C.L. 0·472 in.
   Initial setting size 0·469 in.
   Lower P.C.L. 0·458 in.
   Drawing lower limit 0·455 in.
   480 pieces may be expected to be machined between resets.
3. Approximately 25 pieces machined per 0·001 in. drift.
   Standard deviation 0·0004 in.
   Drawing upper limit 0·2520 in.
   Upper P.C.L. 0·2514 in.
   Initial setting size 0·2492 in.
   Lower P.C.L. 0·2486 in.
   Drawing lower limit 0·2480 in.
   40 pieces may be expected to be machined between re-sets.

Chapter 8

1. Upper ($\frac{1}{1000}$) Action limit for sample mean = 0·4210 in.
   Upper ($\frac{1}{40}$) Warning limit for sample mean = 0·4193 in.
   Lower ($\frac{1}{40}$) Warning limit for sample mean = 0·4139 in.
   Lower ($\frac{1}{1000}$) Action limit for sample mean = 0·4122 in.

Upper ($\frac{1}{20}$) limit for sample range = 0·0123 in. σ = 0·00318 in.
Proportion expected outside limits = 1·9 per cent.
Minimum proportion outside limits = 0·17 per cent, achieved by shifting process mean to 0·420 in.

2. Upper ($\frac{1}{1000}$) Action limit for sample mean = 0·1000+0·0011 in.
Upper ($\frac{1}{40}$) Warning limit for sample mean = 0·1000+0·0007 in.
Lower ($\frac{1}{40}$) Warning limit for sample mean = 0·1000−0·0008 in.
Lower ($\frac{1}{1000}$) Action limit for sample mean = 0·1000−0·0013 in.
Upper ($\frac{1}{20}$) limit for sample range = 34 × 0·0001 in.
$\bar{X}$ = 0·1000−0·000 068 in.
$\bar{w}$ = 20·5 × 0·0001 in.
3. Limits for process mean ($\bar{X}_u$ and $\bar{X}_L$) = 0·1000 ± 0·0024 in.
Warning limits ($\frac{1}{40}$) for sample mean = 0·1000 ± 0·0031 in.
Action limits ($\frac{1}{1000}$) for sample mean = 0·1000 ± 0·0036 in.
σ = 8·8 in. × 0·0001.
4. Upper ($\frac{1}{1000}$) Action limit for sample mean = 0·4371 in.
Upper ($\frac{1}{40}$) Warning limit for sample mean = 0·4359 in.
Lower ($\frac{1}{40}$) Warning limit for sample mean = 0·4316 in.
Lower ($\frac{1}{1000}$) Action limit for sample mean = 0·4304 in.
Upper ($\frac{1}{20}$) limit for sample range = 0·01065 in.
Samples 1–10 inclusive: $\bar{X}$ = 0·43378 in. $\bar{w}$ = 0·0067 in. σ = 0·00264 in.
8½ per cent of output expected to be outside limits.
5. (a)

	Mean in.	Standard deviation in. × 0·001
Head thickness	0·1239	0·94
Overall length	0·5293	1·18
Shank diameter	0·1855	0·41

(b) Mean head thickness should be reduced by about 0·004 in.
Mean shank diameter should be reduced by about 0·0005 in. to minimize proportion of non-conforming work produced.

(c)

Limits for Sample Mean	Thickness (Inches)	Length (Inches)	Diameter (Inches)
$\frac{1}{1000}$ Upper action	0·1236	0·5383	0·1856
$\frac{1}{40}$ Upper warning	0·1231	0·5377	0·1854
$\frac{1}{40}$ Lower warning	0·1169	0·5173	0·1846
$\frac{1}{1000}$ Lower action	0·1164	0·5167	0·1844
$\frac{1}{20}$ Upper limit for range	0·0034	0·0043	0·0015

Chapter 9

1. $\bar{p}$ = 0·073
$\bar{n}$ = 741·6

$\sqrt{\frac{\bar{p}\bar{q}}{\bar{n}}}$ = 0·009 553

$\frac{1}{1000}$ Upper limit for p = 0·1025
$\frac{1}{40}$ Upper limit for p = 0·0917

Weeks 2 and 4: $p >$ action limit. Revision of limit using actual n puts them further out, since $n > \bar{n}$ and limit value *reduced.*

2. $\bar{p} = 0{\cdot}0314$

n	120	140	160	180	200
$\sqrt{\dfrac{\bar{p}\bar{q}}{n}}$	0·01592	0·01473	0·01378	0·01299	0·01233
$\frac{1}{1000}$ Upper limit for p	0·0806	0·0769	0·0739	0·0715	0·0695
$\frac{1}{40}$ Upper limit for p	0·0626	0·0602	0·0584	0·0568	0·0555

Sample 2: $p >$ Warning limit.

3. $\bar{d} = 9{\cdot}95.$
   $\frac{1}{1000}$ Upper limit for $d = 19{\cdot}7.$
   $\frac{1}{40}$ Upper limit for $d = 16{\cdot}1.$
   Sample 2: $d >$ Warning limit.
4. $\bar{d} = 2.$
   $\frac{1}{1000}$ Upper limit for $d = 6{\cdot}2$ *or* $\bar{p} = 0{\cdot}05.$
   $\frac{1}{40}$ Upper limit for $d = 4{\cdot}8.$
   $\frac{1}{1000}$ Upper limit for $p = 0{\cdot}153.$
   $\frac{1}{40}$ Upper limit for $p = 0{\cdot}119.$

N.B. $40 \times$ limits for p give 4·76 and 6·14, i.e., good agreement with limits for d.

5. $\bar{p} = 0{\cdot}08$ for lots 1–10 inclusive. Limits for p, based on $\bar{n} = 147{\cdot}7$, are:
   $\frac{1}{40}$ Upper warning limit: 0·124.
   $\frac{1}{1000}$ Upper action limit: 0·149.

   Lot 16. $p = 0{\cdot}151.$
   Outside *average* action limit.
   Revised action limit for $n = 126$ is $p = 0{\cdot}155$, i.e., inside *corrected* action limit.

   Lot 28. $p = 0{\cdot}150.$
   Just outside average action limit.
   $n = 175$ (i.e., $n > \bar{n}$); corrected limit will be *lower* and p remains above action limit.

Chapter 10

1. No.
2. No.
3. No.
4. No.
5. Yes.
6. (a) $\mu = 0{\cdot}824 \pm 0{\cdot}006$ in.
   (b) $\mu = 0{\cdot}824 \pm 0{\cdot}008$ in.
7. (a) 246 (b) 426
8. $F = 1{\cdot}33 \quad v_1 = 20 \quad v_2 = 12.$
   Not significant at 5 per cent level.
   Yes.

Chapter 11

1. $\chi^2 = 7{\cdot}0$ with 4 degrees of freedom.
   Not significant at 5 per cent level.
   No reason to doubt that all compressors are equally good.

2. $\chi^2 = 5{\cdot}0$ with 1 degree of freedom.
   Significant at 5 per cent level but not at 1 per cent level.
   There is some reason to doubt that line A is as good as line B.
   $\chi^2 = 12{\cdot}8$ with 1 degree of freedom.
   Significant at 0·1 per cent level.
   There is very strong reason to doubt that line A is as good as line B.
3. $\chi^2 = 25{\cdot}0$ with 4 degrees of freedom.
   Significant at 0·1 per cent level.
   Strong reason to doubt that failure rate is independent of wattage. Inspection suggests that failure rate is lower for 15W bulbs than others.
4. $\chi^2 = 9{\cdot}5$ with 4 degrees of freedom.
   Not significant at 5 per cent level.
   No reason to doubt that all 5 moulds are equally good.
5. $\chi^2 = 2{\cdot}2$ with 4 degrees of freedom.
   Not significant at 5 per cent level.
   No reason to doubt that extrusion quality is independent of pressure used, over the range tested.

Revision examples

1. (a)

p	0·03	0·07
Probability of 1 defective in sample of 20	0·3364	0·3526

   It is more likely that the batch contains 7 per cent of defectives.

   (b)

p	0·03	0·07
Probability of 1 defective in 20 and 2 in 50	0·0859	0·0650

   It is more likely that the batch contains 3 per cent of defectives.

2. Poisson approximation to Binomial.
   $\lambda = np \qquad \lambda_1 = 1 \qquad \lambda_2 = 5$
   $P(>2) = 1-P(\leqslant 2)$ (a) $P_1(>2) = 0{\cdot}0803$
   (b) $P_2(>2) = 0{\cdot}8754$.
3. $P = (1-p)^{49}(1+49p)$ P.R. = 0·2642 C.R. = 0·1900
4. $P = P_{(0)}+P_{(1)}[P'_{(0)}+P'_{(1)}]+P_{(2)}P'_{(0)}$
   where
   $P_{(0)} = (1-p)^{30}$ $P'_{(0)} = (1-p)^{20}$
   $P_{(1)} = 30p(1-p)^{29}$ $P'_{(1)} = 20p(1-p)^{19}$
   $P_{(2)} = 435p^2(1-p)^{28}$
   (a) 0·0746 (b) 0·4342
   (c) A.O.Q.: (i) 0·0185, (ii) 0·0261
   $\bar{n}$: (i) 38·7, (ii) 41·5.
5. Frequencies 5, 10, 16, 45, 10, 9, 5.
   (c) 7·264 mm (d) 7·255 – 7·275⁻ mm
   (e) 0·0125 mm (f) 7·2634 mm
   (g) 0·02766 mm (h) 0·38 per cent
   81 per cent expected to conform to specification.
6. $\bar{d} = 1{\cdot}7$.
   Upper warning limit = 4·3
   Upper action limit = 5·6
7. $\bar{p} = 0{\cdot}0108 \quad \bar{n} = 250$ (approximately constant)
   Upper warning limit = 0·0239
   Upper action limit = 0·0304

8. Process mean = $+0{\cdot}165$ watt above nominal
Mean range = 1·07 watts
Upper ($\frac{1}{20}$) limit for sample range = 1·88 watts
Warning limits for sample mean = $0{\cdot}165 \pm 0{\cdot}514$
Action limits for sample mean = $0{\cdot}165 \pm 0{\cdot}803$
watts difference from nominal
9. 12,856 g to 12,469 g.
10. (a) Unpaired samples.
$\bar{x}_A - \bar{x}_B = 0{\cdot}0116$. $t = 0{\cdot}35$
22 degrees of freedom. Not significant at 5 per cent level.
(b) Paired samples.
$\bar{d} = 0{\cdot}00583$ $t = 0{\cdot}94$
11 degrees of freedom. Not significant at 5 per cent level.
In neither case does the evidence give cause to doubt that the two gauges read the same.
11. $F = 1{\cdot}14$ $v_1 = 8$ $v_2 = 8$
Not significant at 5 per cent level, so there is no reason to doubt that the plate thicknesses are equally variable.
12. $\bar{x}_2 - \bar{x}_1 = 0{\cdot}002$
n is large, therefore 't' tends to Normal.
$$Z = 6{\cdot}25$$
Very highly significant. Unlikely that daily concentration has not changed.
13. σ known, therefore Normal variate.
$Z = 3{\cdot}77$.
Highly significant, i.e., strong reason to suppose it was not an ordinary day.
14. μ known. σ unknown.
$t = 2{\cdot}18$. $v = 9$.
Not significant at 5 per cent level.
No reason to reject manufacturer's claim.
15. $\lambda = 0{\cdot}8$ breakdowns per day.
'Expected' frequencies: 89·9, 71·9, 28·8, 7·7, 1·5, 0·2.
$\chi^2 = 3{\cdot}66$. $v = 3$.
Not significant at 5 per cent level.
Accept hypothesis that distribution is Poissonian.

Index

Scylla Rider

Singing Kettle

KOP19